Architects and their Practices

Architects and their Practices

A changing profession

Martin Symes
School of Architecture, University of Manchester
Joanna Eley
School of Architecture, University of Manchester
Andrew D. Seidel
College of Architecture, Texas A & M University

BUTTERWORTH
ARCHITECTURE

Butterworth Architecture
An imprint of Butterworth-Heinemann Ltd
Linacre House, Jordan Hill, Oxford OX2 8DP

ℛ A member of the Reed Elsevier group

OXFORD LONDON BOSTON
MUNICH NEW DELHI SINGAPORE SYDNEY
TOKYO TORONTO WELLINGTON

First published 1995

©Martin Symes, Joanna Eley and Andrew D. Seidel

British Library Cataloguing in Publication Data

A catalogue record for this book is available from the British Library

ISBN 0 7506 1299 1

100055279 X

Library of Congress Cataloging in Publication Data

A catalogue record for this book is available from the Library of Congress

Produced by Graham Douglas, Bath
Printed and bound in Great Britain

Contents

Preface

In 1968 (the year of the Paris uprisings) Lords Esher and Llewelyn-Davies published a review of changes which might affect the architectural profession by 1988.[1] They foresaw increasing integration of the construction industry, a growth of specialization within design teams, major changes in architectural education and great uncertainty over the role of the professional institute. While broadly correct, their predictions failed to identify the extent to which architects would lose their role in planning for the welfare state, or how far their practices would be exposed to the booms and busts of a market economy.

This book sets out to explore what might be considered a test case of the changes which have hit many professions in the last two decades. It has developed out of a research project funded by the Architects' Registration Council of the United Kingdom and a Wolfson Foundation grant administered by Manchester University. Initially this consisted of seven case studies, covering a wide range of practice types. The same methodology was used for all seven studies, and since together they present a comprehensive picture of the problems and achievements of contemporary practice in 1991, all are discussed in this book. A selection of projects undertaken by these firms is also described.

Some time later the authors initiated a wide-ranging questionnaire survey of the perceptions of practice by members of the Royal Institute of British Architects and this is reported in detail here. Involvement in a parallel research programme into the future role of the architectural professions, being undertaken by the French Ministère du Logement, Plan Construction et Architecture and frequent contact with research groups in the United States have given us the chance of broadening our thinking about the possibilities for the future. In the meantime the Royal Institute had begun to publish studies on this topic.[2] The Government stimulated discussion on the future of the Registration Council, and a further debate was initiated on the funding and content of architectural education. These two events added considerable relevance to the work we had done.

A brief discussion of the early stages of our investigations has already been published,[3] and progress statements have been made at various conferences.[4] Various details have been published in University departmental discussion papers.[5] Butterworth Architecture then agreed to publish a book which would seek to introduce the theoretical issues raised by attempting to study the evolution of a profession and to extend architects' understanding of the nature of change in the practice of their profession.

Some key aspects of the profession's response to change are discussed in Part One. Part Two reports our questionnaire survey, covering the views of a statistically significant cross-section of partners or principals of RIBA firms. Followers of the recent debate might expect it to show what specializations had developed in British architectural practice during the boom, when demand was expanding, and which seem to be surviving in the recession. They might also expect to learn how much new technology has been introduced. In addition, however, they may be either pleased or surprised to find that the management tasks undertaken by architects are a great deal more wide ranging than has hitherto been appreciated. There is a wealth of new information here and we believe it merits detailed study. The case studies in Part Three aim to show how a variety of architectural design firms have developed in the last two decades, and how the role changes involved have been viewed by some of the individual architects concerned. Part Four brings some of these observations together in the context of specific projects, drawn from the experience of the case study firms. Part Five provides an opportunity for opening up a discussion on future trends.

In addition to thanking our publisher's team we must therefore express our deepest gratitude to the Registration Council for their generosity in providing early funding for our developing studies. The confidence they have implicitly shown in the importance of this type of work has been fundamental. We hope we have added value to their initial grant.

Equally fundamental has been the generosity of the many hundreds of architects who responded to our questionnaire and of the seven firms whose practice histories are described in the case studies. Without their interest, patience and co-operation there would have been no study. They have committed resources of many kinds to the project and are, we believe, as clear as we are that the profession can only benefit from its talent, dedication, imagination and flexibility being better understood by the public at large. We hope we can show our gratitude to them by making a convincing contribution towards this important change of perception.

As this work has originated in schools of architecture, a brief discussion of the academic viewpoint and methodologies which have been adopted is perhaps to be expected. It may show that attempts have been made to ensure that the work has value beyond its immediate interest to those architects struggling with the problems of earning a living in today's harsh economic climate and that students can learn from it some more general lessons concerning the forces for change in the profession for which they are preparing themselves.

The purpose of the present study is thus to give as objective a picture as possible of the transformations which seem to be taking place within the architectural profession and then to let a judgement concerning any problems the profession is experiencing emerge from that picture. None the less it is clear that at certain points in the study decisions have had to be made about the way information would be collected and reported. Should we represent the profession descriptively, as it seemed to be, or prescriptively, as it should or could be? The latter might have been preferred if the study's first aim were only to support the profession's claims to power and prestige. But the former approach would appeal to a more critical mind and it is that which has been selected whenever possible here.

There are a number of academic precedents for studies of a professional group. The main thrusts that such studies have taken in the past can be briefly reviewed.[6] A first set might be considered as functionalist. They use the group's own definition of its membership, study the activities of those members and their ways of organizing themselves, and then discuss the function that this type of specialized pattern of work and associated behaviour may have for the group's position in society. A second set might be considered as critical. These begin by taking a point of view outside the situation being studied. Typically it is assumed that there are forces which provide a structure to society and have an impact on a particular kind of work or the lives of those who undertake it. Ultimately such studies attempt to show that the particular form of institutions a group develops must be constrained by their position in the larger pattern. Both functionalist and critical perspectives have influenced the research programme being reported here.

The case studies of offices and of projects were completed first. They involved visiting seven offices, meeting their principals and other staff, reviewing the history of their practices, studying a number of their projects and questioning them about their experiences. To make sense of these cases a 'critical' perspective was taken; indeed to some extent it had to be. We wanted to see how architects were compelled by economic forces both to diversify and to engage in potentially destructive competition with the other actors in the building development process. We wanted to see how the process of project development and design had been affected by such changes of context.

The questionnaire survey, however, was begun and completed later. It covered a sample of the partners or principals of British architectural practices, obtaining, by post, information on these architects' behaviour and attitudes. This gives a comprehensive overview of the way this level of the profession performs and of the roles its members play, not only in building design but also in technological development, in construction management and, increasingly perhaps, in identifying the financial value to be extracted from the environment. It shows how they 'function'.

Taking both research perspectives together, we saw architects as having a strong and surprisingly consistent world-view. This has helped them to resist many of the pressures to abandon that sense of being a special kind of practical and artistic person which has so often marked them out. We are optimistic about the future but feel architects should develop a more realistic understanding of the services their clients demand and of the managerial tasks they will be called on to perform. Only in this way will they continue developing their contribution to the quality of environmental design.

Notes

1. Esher and Llewellyn-Davies, The architect in 1988. *Journal of the Royal Institute of British Architecture*, Vol. 75, No. 10, 1968, pp. 448–55.

2. See the findings of the First Phase of the *Strategic Study of the Profession*, London, RIBA, 1992.

3. In Barrett, P. and Males, R. (eds) *Practice Management*, London, Spon, 1991.

4. At the 11th and 12th Conferences of The International Association for People–Environment Studies, held respectively at Ankara in 1990 and at Thessaloníki in 1992, at the People and Physical Environment Research conference in Australia at the Architecture Research Centre's meeting in Washington in 1993, at the 1993 conference of the Environmental Design Research Association in Kansas and in the same organization's 1995 conference in Texas.

5. Eley, J, and Symes, M. (1993). Recent changes in architectural practice; occasional papers in *Architecture and Urban Design*, No. 2, University of Manchester School of Architecture, 1993; and Seidel A. D. and Aravot, I. (1994) The lost architect's desk reference and the need for rewriting it. In: *Magazine of the Faculty of Architecture and Town Planning*, Haifa, Israel Institute of Technology, 1994.

6. For a more complete discussion readers might like to refer to Friedson, E. The theory of the professions: state of the art, in Dingwall, R. and Lewis, P. (eds) *The Sociology of the Professions*, Basingstoke, Macmillan, 1983; and Larson, M.S. *The Rise of Professionalism: A Sociological Analysis*, Berkeley, University of California Press, 1977. They refer to the seminal work of Parsons, Hughes and Friedson in defining these different approaches to the study of the professions.

Part One

Architecture
as a
Profession

1

Economic and social factors

Challenges to the architectural profession

Architects seem to live in a world of their own. Membership of the profession
has been closed. Methods of work have not often been divulged to building
users or shared with collaborators in the building industry. Architects have
had unique methods of rewarding success; their working environment has been
highly specialized; and they have developed a number of private languages for
the discussion of building quality. This culture has been challenged in the last
twenty years. Economic recession, social protests and media exposure have
threatened its self-sufficiency.

The RIBA study mentioned in the Preface has attempted to contrast the
advice given to architects establishing or modernizing their firms in the 1960s
with a picture of the restructured profession in which the equivalent group
would be expected to operate today.[1] In the earlier period it was imagined that
the ideal firm would employ a total of up to 25 people, including secretarial
staff and some technicians. It would aim to be competent, or to be able to
develop competence quickly when necessary, over a wide range of building
types. It would take over the task of organizing the procurement of built space
for the non-expert client, engage specialist consultants as required and
supervise the construction process.

Architects working in such an environment would expect to follow a building
design through all stages of its development and implementation and to
understand how to balance at various stages the needs of building quality and
functional performance against pressures of cost and time control. In effect
they were to be general managers of the provision of a new and improved
environment. If many of them were to find themselves working in a government department this was not likely to alter the picture. Governments were

3

also providers of this better environment and the principles of organization which would maximize the benefits to be gained from designers' participation in the process still applied.

At the present time the profession must admit to a very different situation. Many architects find themselves working as consultants to large business organizations or, which is functionally rather similar, as contract workers for architectural entrepreneurs. Others, as elsewhere in Europe, work in small one- or two-person businesses taking a variety of types of commission inside and outside architecture proper.

Those who work in the public sector are rarely now the organizers of a building design. More likely they act as some kind of regulator of the public interest, looking after health and safety or financial matters.

Design proposals are now typically passed from one architect to another along the chain from concept to realization. Judgements of quality and of value for money are made by specialists in various aspects of management science. The architect may still have been trained as a generalist designer but now almost inevitably finds himself or herself working in a more specialized role. Those who concentrate on management find it is the co-ordination of these specialists, rather than the overall quality of the environment that results, which takes up their energy and time.

The survey, case studies, and projects reported in this book explore the changes which have been wrought in more detail, focusing both on the way architects' work is being re-organized and on the personal contributions individual architects make to their collective task of constructing our environment.

The data in the book are intended to throw light on the roles of the architect, the characteristics which architects hold in common, the economic factors which have led to the emergence of new ideas about what an architect is and does, and the choices which are now emerging for important subgroups within the profession. Finally the book hazards a few guesses as to how these choices may be made as we approach a new century.

The development of the profession

The starting point for this investigation is that architects define themselves as professionals. They claim financial rewards for their knowledge and skill in the design of built environments. In addition they claim the respect due to those who protect the public interest and the needs of those who do not possess this knowledge and skill.

Unlike other professional groups, however, they also claim to contribute to the artistic culture of their country. The fortunes of such a professional community in times of economic, social and cultural change are of significance not only for the quality of the environment itself but also for society's perception of the value of encouraging other specialist groups to make similar claims.

An important study[2] has suggested that the architectural profession has placed, or been encouraged to place, emphasis on different dimensions of its professionalism at different periods in its history. In the nineteenth century it was common for architects to think of themselves primarily as artists. They worked as individuals whenever they could. They received their commissions through patronage from sophisticated amateurs and resisted pressures to form themselves into a professional or trade association, preferring the establishment of learned societies or academies in which the finer points of their art could be debated.

In the earlier part of the present century, both in Britain and in North America, a more gentlemanly concept seems to have developed. Architects established themselves in a privileged position within the expanding middle classes, taking responsibility for the design and construction supervision required for the accommodation of their social peers in efficient and sanitary surroundings at home and at work.

It was in this period that architecture became more clearly professionalized: institutes were set up to which architects were by law required to belong; and training was organized, first through apprenticeships and later through the setting up of schools awarding diplomas or university degrees. Increasingly, architects took salaried employment, rather than relying on fees for their income, and a growing proportion joined government service, either to assist in the implementation of social programmes or to control the standards of work done by others.

Since the Second World War a new development has been the emergence of architect-entrepreneurs. The building design team, normally until very recently led by an architect, has come to see itself as a business organization providing services in a market-place. The owners of these businesses have needed to establish a product line, to assemble capital – human, material or financial, to promote their output, and to evaluate their profitability.

No historical sequence of such a simple kind can fully describe the complexity with which social and economic developments occur. The broad picture could be filled out with details of overlapping trends, mixed motives and confusing changes of direction. So the author of the study mentioned above suggests that although these types of architect may have seemed to predominate in particular periods, all three aspects of the architect's role have usually been present to some degree: it is the balance or emphasis which has changed. So today, when change seems to be in the air again, and new opportunities

5

seem to be opening up, it may be appropriate to suggest that it is the mix of artistic, community service and business aims which is being restructured. If it is in such a context that the present differentiation of architects' roles is taking place then there may be options available to the profession and real choices which can be made.

The need for specialization

It is surely fair to assume that the first cause of these changes in architectural practice is economic and that the institutional developments and personal problems, which have also been sketched, probably follow from this.

It is worth recalling that the business cycle of the British economy has experienced two booms in the post-war period, one in the late 1960s and the other in the 1980s. It has also suffered two busts, one associated with the oil shock of the early 1970s and the other with the world-wide recession of the early 1990s.

The first two questions for this book to answer – the response to economic change and the adoption of new technology – are connected with the development of specialization, or in the words of one of the grandfathers of modern economic thought, John Stuart Mill, the emergence of a new division of labour. One cause of specialization was identified by Mill as economic growth, or:

> ...the level of demand. As [this] rises it becomes possible to increase the degree of specialization of production at each stage...The classic example was the cotton industry. As demand rose the interdependent stages of production: spinning, weaving and finishing, became detached and separate producing units were established.[3]

We might then expect that when times are hard some of these specializations will find themselves less viable and will die away. Their tasks will be picked up by those undertaking other, more robust, areas of work. It will be as if 'labour', once divided, is later forced to recombine, but possibly in new and different groupings. Naturally these form the foundation for the expansion of the next boom. So if we look carefully at the structure of an industry in the throes of a recession we may, by thinking of where growth will next occur, be able to speculate on the shape which that industry will have when the next boom arrives. There may be new kinds of firms doing different mixes of work as well as new divisions of related industries that include in their work various tasks which they had previously left to others.

Our study does suggest that architecture had begun to develop important specializations[4] in the last upswing of demand. It may also suggest that firms in some of these specializations have gradually moved into collaboration with some of their competitors during the present recession. The architectural profession which eventually grows again when the economy expands may thus be more specialized but in different areas than before. Some functions may even be lost to the profession for ever.

An American study[5] suggests that in that country a number of interesting new sub-professions have begun to emerge. There are 'architect-stars' who appear on television and tour the nation's lecture halls without necessarily producing buildings; there are architectural artists who sell their drawings and project ideas; there are architect-developers, architect-engineers, building construction experts and so on, each group with its own relatively independent career path. In this country, the Prince of Wales has made the general public aware of community architects and of those who preach classicism as if it were a religion. In some schools of architecture, students are already being offered optional studies in such areas and are thus being drawn away from the traditional mainstream at an early age.

Mill's other cause of the division of labour is to be found in technological development. Changes in the technology available tend to be harnessed to the establishment of new specializations, and to reinforce their economic position within the relevant industry. So far as these observations about the introduction of new technology are concerned, in architectural design we may not have to look far to observe the process. Computer-aided design (CAD) seems to have spread like wildfire in good times and bad. Architects, and others, will be interested to see how far this development reinforces the other changes which are taking place in the structure of the develop–design–build process.

The sizes of firms

Perhaps we should point out that architects are not alone in experiencing structural change in recent years. This has undoubtedly been occurring in all the service industries. The trend began in the late 1960s. In the words of one rather critical commentator, we have experienced since then the 'complete reorganization of the global financial system',[6] with the emergence of, on the one hand, massive conglomerate organizations and, on the other, great pressures to deregulate the economic system. In addition it is clear that during this period access to specialized information of all kinds, including that of the professional building designer, has become increasingly valuable to managers and decision-makers. So the same text also says that:

In a world of quick-changing tastes and needs and flexible production systems... access to the latest technique, the latest product, the latest scientific discovery implies the possibility of seizing an important competitive advantage.

Structural change can be clarified by a short excursion into economics. The combination of small businesses into larger ones is a kind of economic integration. Of this two main types are possible: vertical and horizontal.

Horizontal integration is a strategy for increasing market domination in times of potential growth. Here specialists who work in parallel at the same stage of production get together so that 'the cost [is reduced] or the quality of the product... is superior to what could possibly be achieved by operating independently... [and] synergy of various forms may exist.'[7] In building design the 'one-stop shop' of multi-disciplinary practice gains exactly these advantages.

Vertical integration, on the other hand, is achieved when the earlier and later stages in production are brought under the same ownership. Design–build is one of the results of this type of integration in construction:[8] the builder offers design services to the purchaser and the latter need not employ a consultant as intermediary. According to some economists this kind of integration does not have specific benefits for profitability, that is it 'does not increase the percentage of the market controlled by a firm', but it does allow better business planning and 'can be regarded as a substitute for a long-term contract [and] ensure a stable input flow at some future date.'[9] Thus the uncertainty created, for builders and designers, of working with different teams on each separate project is removed.

Thus vertical integration is probably best thought of as a defensive mechanism for when times are bad and prospects are bleak, as indeed they have recently been in the building industry, whereas horizontal integration is an attacking mechanism for periods of expansion. The last twenty years have seen an alternation of both these economic contexts and the changes which have taken place in building design result from a combination of both types of integration.

The tendency for the continuing growth of professional firms in size and complexity has been encouraged politically by the vaunting of the virtues of a private sector dominated market-place for professional services as well as by an anti-institutional 'backlash' of antagonism to state intervention in design and development.[10] Perhaps unsurprisingly it was the public sector which experimented most with vertical integration. The schools, hospital and housing programmes of the post-war period created strong ties between research, development, briefing, design, construction and maintenance, which were often carried out by the same authority. Privatization has split these activities off from each other at just the time when technical complexity has increased in the doing of many of them. Where horizontal integration has been possible it has been the large private firms which have taken advantage of it. The current revival of vertical integration through private sector design–build may well not survive the ending of the recession and whatever building boom eventually emerges.

Types of practice

The third and fourth questions for the book to address are thus the changes taking place in the structure of architectural practices, in particular the growth of large private firms, and the consequences of the decline of the public sector.

Architects now face increasing competition for the provision of the more specialized services such as project management; they face increasingly difficult choices as to whether they should remain small one- or two-person firms or join large multi-disciplinary practices; and they face increasing disillusion with their ability to promote and protect the public interest by design. One argument frequently heard, but not supported by our data, is that the profession is dividing into a dual economy of small, high-quality, design-led entrepreneurs and large, efficient but mediocre business-oriented companies. In this evolution, it is argued, both the medium-sized all-rounders and the committed public service organizations such as local authority housing design divisions must eventually disappear. [11]

There are many interesting facets to this kind of change. Does the difference between large and small firms reflect a growing difference between the commercial and artistic sides of practice in architecture? Are the architects who work in large organizations taking advantage in some way of their membership of multi-skilled businesses or losing out in the competition for certain kinds of work? Do the many architects who continue to operate in tiny firms fear competition from other professions, or trades, and are some of them able to establish a niche on grounds of special knowledge or a peculiar talent? Are there ladders of opportunity which lead practitioners or their businesses from the small to the large?

Of particular interest, clearly, is the question as to whether the ideal medium-sized practice much vaunted by the RIBA in the 1960s can still be identified in the 1990s. The one we studied as one of our cases has faced a number of commercial difficulties and seemed as we wrote to be oscillating between a programme of organizational growth and one of contraction. Even to talk of growth with internal subdivision (a possible way of squaring the circle) is to suggest that for each subgroup a smaller size of working group is to be preferred, but within an industrial situation which requires large-scale organization. Is a new role for the architect being forged, it would seem, as one of privilege within a team rather than as leader of it?

A rather special set of questions is raised by the decline of the public sector. Many building types were almost exclusively the province of government or local government architects in the heyday of the welfare state. Does the fact that so much of this work is now undertaken both by and for private organizations mean that architects will specialize less, rather than more, in these areas? How has the research data assembled by groups concerned with schools, housing, hospitals and universities been distributed around the profession? If there is a new type of public architect – the regulator – then what sort of work does he or she do? Again, the question arises, can one move into or out of these roles or is a specialization, once established, a specialization for life? The survey provides some answers to this sort of question and the case studies some others.

This discussion brings us back to the broader problems which underlie our research programme and give it continuing relevance. On the one hand we are asking whether these changes are all for the good; and on the other hand we are asking whether the institutions of the architectural profession are responding adequately to the needs of firms and individuals in these evolving situations.

The values and the institutions

These questions are not only practical but also moral. As stated earlier, although their special skills undoubtedly have a monetary value in the marketplace, professionals claim other privileges because they can represent higher values than are usually found there. How these can be expressed, and indeed justified, is a question which has concerned sociologists for many years. A recent commentary explains how one of them, Talcott Parsons, 'was committed to the moral autonomy of the individual... [seeing] unrestrained... capitalism as a threat [which he] hoped could be defended... [through] the underlying values of professionalism'.[12] From this point of view, then, professionals must maintain the freedom to think and say what they believe to be right, but economic forces will tend to undermine them. We should consider whether this has been happening to architects in the present period of rapid change.

It is the argument of occupational sociology that our social institutions have arisen to fulfil the function of defending social values. Some 90 years ago one of the founders of this discipline, Émile Durkheim, suggested that social problems are caused by the division of labour into specialized groups and by technological advances.[13] These can create such stress that they must be counteracted by the development of strong occupational associations.[14]

So the fifth and sixth main questions for our study to address are how this profession still manages its collective affairs, and how its members cope with changes to their personal involvement. The loss of a sense of involvement with one's work, which comes from only taking on a particular part of it, may become more tolerable if one can belong to a community of workers with similar tasks. In addition a number of such communities can cohere again by taking on, together, all parts of the work. This togetherness Durkheim called 'organic solidarity'. He argued that it was becoming the characteristic form of modern society. A study of a professional group such as the one reported on here may give some insight into the relevance of such an analysis today. Our study of the way architects have responded to change may allow us to understand better the ways in which

> ...between [society] and the individual, there is intercalated a whole series of secondary groups near enough to the individuals to attract them strongly in their sphere of action and drag them... into the general torrent of social life... [as well as]... how occupational groups... fill this role.[15]

Architects have professional bodies, of which the Royal Institute of British Architects is the best known in this country; quasi-governmental organizations, as the Architects' Registration Council might be described; European Directives, an International Union, schools of architecture, regional associations and so on. Architects' working lives are thus organized into a pattern and protected to a degree from some of the effects of economic forces. But as the division of labour develops, these institutions are surely developing as well.

Architects' personal achievements are recognized by competition successes, professional awards and national honours. A sense of separate identity is created and reinforced by books, journals, magazines. But more than this, architects often live together, choosing the same towns, suburbs and villages, buying or renting similar houses, shopping at the same stores or even supermarkets. They and their companions or families drink in certain pubs, holiday on the same coasts or in the same mountain resorts. These forms of association reinforce our perception of architects as having a special place in our culture.

At one level it is this sense of being a coherent group within an organic society which is threatened by the changes that are taking place. To allow free competition, to encourage advertising in the press, to promote participation in non-professional business ventures, to remove the exclusive right to the use of the title of architect, to recognize different forms of professional education: all these changes risk driving architects apart from each other. Differentiation and diversification may well bring adaptability and flexibility with them, but they also weaken the sense of collective identity architects have had and make it less easy for them to combine in playing a particular role in the maintenance of our culture.

The case studies and questionnaire give some answers to questions such as this. They show what institutions architects belong to, what meetings people go to, how they value various kinds of education, as well as whether they still form a coherent community. Of course, some answers are somewhat subjective and thus harder to find, but they are none the less interesting for that. Our conclusions here depend more on inferences made from the circumstances of practice than on specific facts about it.

The idea that a certain kind of working life can lead to a particular form of personal life, and then perhaps be reinforced by it, is a central characteristic of the working community. One of the classic social science studies of the relationship between different segments of daily life used the architectural profession as a case study.[16] It argued that the efforts they made to emphasize the special status of their profession played a central role in the way architects organized their domestic behaviour:

> [A] *strong convergence of the worlds of work and non-work is also displayed... in three respects... relationships, values, and culture and identity. Indeed the very segregation of* [this] *occupation... from other members of society forms the central theme in the shared world-view and perspective.*[17]

The locational aspect of a professional community is also important to understand. Indeed it is linked to that of personal life-style. This comes out particularly strongly in respect of our case study firms. Most of the respondents were known by the researchers and a number of them also knew each other and met quite frequently. It is one of those clichéd 'small worlds' which we are able to portray and it does not seem to have fallen apart yet. Here the role of educational experience is of considerable interest.[18] It provides much more than access to skills: it provides the introduction to a work-centred social life and even, it could be argued, a continuing set of personal values.

The questionnaire survey helps give a wider picture of the way the architectural community continues to live the same kind of life in the same kind of places. We can suggest, despite our critical viewpoint, that social forces have not entirely undermined the special position architects hold in civil society. Of course the book does not delve into those architectural life-styles portrayed in glamorous advertisements or intellectual Italian films, nor does it undertake a detailed geographical survey, but some of the information gathered gives us clues as to how the general problem is being answered.

Notes

1. *Strategic Study of the Profession*, London, RIBA, 1992.
2. Saint, A. *The Image of the Architect*, London, Yale UP, 1983.
3. See McCormick, B.J., Kitchin, P. D., Marshall, G. P., Sampson, A. A., Sedgwick, R. *Introducing Economics*, Harmondsworth, Penguin, 1974.
4. Some of these are reported in the *Strategic Study.*
5. Gutman, R. *Architectural Practice: A Critical View*, Princeton Architectural Press, 1983.
6. Harvey, D. *The Condition of Post-Modernity*, Oxford, Blackwell, 1989, p. 160.
7. Weston, J.F. 'Conglomerate firms', in Yamey, B.S. (ed.) *Economics of Industrial Structure*, Harmondsworth, Penguin, 1973, p. 310.
8. Yeomans, D. and Steel, M. *Professional Relationships and Technological Change in Great Britain*, report to the Ministere du Logement, Plan Construction et Architecture, 1994, available from the authors at the University of Manchester School of Architecture.
9. Oi, W.Y. and Hunter, A.P. 'A theory of vertical integration in road transport services', in Yamey, *Economics*, pp. 253–63.
10. A 'battle' between 'state professionalism' and 'privatised professionalism' is described in great detail in Perkin, H. *The Rise of Professional Society*, London, Routledge, 1989.
11. An American study which investigates this hypothesis is reported in Blau, J. *Architects and Firms*, Cambridge, MA, MIT Press, 1984.
12. Holton, R. J. and Turner, B. S. *Max Weber on Economy and Society*, London, Routledge, 1993.
13. See Durkheim, E. *The Division of Labour in Society* (G. Simpson, trans.), New York, Free Press, 1933.
14. Yeomans and Steel, *Professional Relationships*, give some interesting historical examples.
15. Durkheim, 'Preface to the Second Edition', *Division of Labour*, p. 28.

16. The study is reported in Salaman, G. *Community and Occupation*, Cambridge, CUP, 1974.

17. Ibid.

18. Only recently has a history of architectural education in Britain been published. See Crinson, M. and Lubbock, J., *Architecture, Art or Profession? Three Hundred Years of Architectural Education in Britain*, Manchester, Manchester University Press, 1994.

2

Debates in the profession

As a preamble to the questionnaire, case studies and projects reported later, we first attempt to review how some of the practical issues which have arisen for the profession have been seen by its observers and critics, commentators and journalists. This review begins to provide answers to the questions raised earlier about changes in architects' core skills, in the technologies of building design, in the relationships between public and private sectors, and in the institutions of the profession.

The early parts of this review show how the recent uncertainty about the status of architecture has been portrayed. One of the main concerns which has been expressed has been the emergence of competition from other professions. A principal focus has been the growth of project management as a separate discipline. This worry has been linked with concerns about new technology, about the public view of architects and about the composition of the design team. We are concerned, therefore, with the transformation in the organization of building design and construction which has taken place since the late 1960s and its relationship with the fortunes of the British economy during this period. Readers may find it helpful to refer to the chronology of events and publications given in Appendix I as they follow the discussion.

The latter part of this chapter charts the perception of three particular problems in more detail by looking at the pages of the profession's official journal. These are: the locus of responsibility for construction supervision; the nature and control of professional ethics; and public criticism of architects' aesthetic judgement. These issues have been ones for which the institutions of the profession have had to formulate legal or practical responses to the economic and social changes which their members have experienced.

The general argument presented by the literature reviewed is that architects were unprepared, either individually or as a group, for the effects of rapid change in the level of demand for building services, that they were not sufficiently skilled in the identification of newly emerging market requirements and the adoption of new technologies, and that they found it hard to cope with the new fluidity in cultural expectations. Some authors even go so far as to suggest that architects have *consistently and over a long period* failed to absorb new technical knowledge into their *modus operandi*. In their view it is this failure which has led to the rise of one new professional specialization in surveying or engineering after another.

Interestingly, the original research presented in later chapters casts more than a little doubt upon this story. It suggests that some organizations and many individuals have in fact adapted remarkably quickly to a turbulent environment and that many of their core skills and fundamental values have remained surprisingly stable. None the less some doubt is thrown on the adequacy of the training and education available to the generality of architects as they adapt to the changing context of practice and this problem must be a major concern for the future.

The general picture is that, over time, a number of new sub-professions, including project management, were first allowed to emerge, then welcomed and then, some time later, feared. To begin with the newcomers were seen as no more than specialists within other professions; then they came to be seen as useful bearers of responsibilities that architects were finding it difficult to carry; and then finally they became competitors for some of the architects' prized professional duties. Our research indicates the ways in which these challenges have been met and the range of institutional, organizational and personal responses which have been adopted in some of the main areas of concern.

The need for management skills

The study undertaken by Austin-Smith and others for the RIBA in 1962 reflected the concerns of the first post-war boom.[1] It dealt first and foremost with the architects' function as designers of new buildings. It asked what sort of work was done in an architect's office and how the efficiency of architects could be raised. Yet its conclusions implied that practice was already changing. It suggested that small jobs should have a different fee scale from large ones; that a sub-profession of architectural technicians was needed; that local authority architects had interests in common which did not concern the private sector; that there could be more research on the time spent on different stages in the design process; and that architectural education should be diversified. Significantly, in view of future developments, it also suggested that new forms of contract could be developed and the code of conduct liberalized.

Most revealingly of all, it estimated that productivity would have to rise by up to 63 per cent over the nine-year period 1961–70. The study 'did not find any evidence that [at that time] management consultants had penetrated far into the complexities of the design process or the organisation of the design group'. The research team found that almost every office they visited was in need of assistance and that 'it should not be impossible to make a [management advice service organized by the RIBA] self-financing in the long run.'

The risks of new technologies

This, the first boom, was a period of growth, and in periods of economic expansion buildings are planned more confidently, created more quickly and built with more innovative technology than in quieter periods. So in 1968 a new Town Planning Act was passed, allowing for strategic thinking as well as mere development control. A record number of new housing unit completions was recorded (414,000) and plans for a totally new Third London Airport were being considered by the Roskill Commission. Expansion was in the air and new technologies, such as large concrete panel construction and sheet asbestos insulation, were introduced partly to meet the demand as rapidly as possible.

A number of unreasonable risks seem to have been taken and the criticisms of poor performance by the architectural profession with which the public is now familiar were just about to emerge. They were perhaps inevitable. None the less what Esher described in 1981 as 'the moral revolution of the seventies'[2] hit architects hard.

The collapse of the Ronan Point housing scheme in East London was thus just one example of a wider malaise, moral as well as economic, which then developed. In 1969 the Skeffington Committee recommended more public participation in town planning, and Colin Buchanan's 1971 note of dissent to the final Roskill Report threw the values of bureaucratic decision-making into question. The moral authority assumed by architects was seen to be lost by 1978 when Watkin[3] published his criticism of the design philosophy of the International Style.

As management consultants were already beginning to make an appearance on the construction industry scene, it was hardly surprising that designers as well as owners and users were pleased to be offered a chance to pass some of the responsibility for risk-taking to them the next time round. The need for this new specialization soon became established.

Much the same story can surely be told of the second boom of the 1980s. The development of London's Docklands which began in 1979 has become a symbol both for the significant expansion of building activity which took place throughout Britain in this period, and for poor quality design. But by this period the professional situation had already been transformed. In 1989 Dowson's phrase 'we were all design consultants by then'[4] provided a neat summary of the changes which had occurred.

The search for new markets

Whereas in periods of expansion all professions, including those concerned with building, may seek to diversify their skills, in a recession they must establish as many new ways of marketing themselves as they can. People coming from quite different backgrounds may find themselves offering to undertake the same work.

Lyall's 1980 review shows how, after the 'official end of the modern movement', the architectural profession departed 'in search of new opportunities'.[5] In 1974 Gordon's study of long-life and loose-fit design laid claim to a new understanding of user requirements.[6] In 1973 Essex County Council's *Design Guide* reassessed the role of architectural design in town planning. The RIBA began to discuss whether architects could advertise in 1977. An even more marketing-oriented approach was taken by the Association of Consultant Architects after its establishment in 1980. The clearly declining economic situation was met by an expansion of the use of design–build contracts, arguably a search for new markets by construction companies as well as by the designers who 'sold' their independence to them. Even Jencks's theories about the new possibilities for architectural expression, published in 1977,[7] may be thought to have supported architects' claims to the attention of different classes of client.

During this period engineers, surveyors and other potential project managers were also not inactive. Having identified building owners' need for expert advice on the repair or refurbishment of the poorer structures of earlier years, they seemed to have found little difficulty in also securing a place in those areas of design briefing and construction site supervision for which architects were losing their credibility. A substantial reshaping of the consultancy side of the building industry was clearly being achieved.

Again, it can surely be argued that a further consolidation and reorganization of the activities of the architectural profession and its existing or potential competitors is taking place in today's recession.

Questions about leadership

This reorganization has had substantial implications for the design professions. In the construction industry architects have not been able to sustain their view that overall control of the building process should be entrusted to only one profession, theirs.

By the early 1990s the RIBA has to have a marketing arm and a list of publications to address the processes of competition. They cover topics such as how to identify potential clients, quality assurance, architects' skills and client needs, and market opportunities for the decade. Today's individual architects see the need to take continuing education courses in marketing, accounting, facility management, real estate development, budget management, office management, project management and construction management together with members of the other professions they meet in the market-place. If architects wish, or need, to establish or retain their claim to expertise in these areas, they will surely find the competition stiff. And indeed, gossip suggests that even more opposition is just round the corner. As internationally known firms of accountants find their core business declining, they are more frequently to be found offering pre-design services to the construction industry and its customers.

Architects may also have lost the argument that building design itself is a single function which must be followed through into the details of a project if the highest levels of quality are to be maintained. Accountability for the quality of building design is now seen to have many facets and be achievable by numerous routes. An anthropologist[8] has said that we do not really know how design quality is achieved. To quote: 'We know very little about what it is like, these days, to live a life centered around, or realized through, a particular sort of ... creative activity.' Clients for building design services who do not understand what is going on will not be easily convinced that large sums of money should be entrusted to single creative individuals. Team-work will be sought and risks will be spread.

The issue in the profession

Thus although the simplest way of defining the architectural profession is that it represents the possessors of a particular set of skills, technical as well as managerial and artistic, the claims made by practitioners to the exercise of such expertise may seem to many surprisingly broad. Architects would like to, and sometimes do, have insight into questions well beyond their professional bailiwick. Indeed technical skills, and the training which is necessary to produce them, are so important to our present state of social development, that they colour all aspects of everyday behaviour. At least this is probably true of the middle classes in Western societies. The likelihood that this form of life will soon spread through the other parts of the world seems hard to deny. Each professional education has its own values to inculcate, each professional task its own way of dealing with economic, political, physical and geographical constraints, each professional institution its own way of responding to change.

In the next section of this review we look at the architects' professional journal to help understand how its particular way of seeing the world has begun to change.

Major areas of change

A number of paradoxes emerge from this discussion.[9] The first is that although architects believe the most important feature of their calling is the opportunity it gives them to exercise individual creativity, they also believe that the conditions of the building industry, in or with which they must work, restrict and inhibit them all. A second paradox probably supports this view. Architectural training is very demanding and it takes a very long time. But architectural practice relies heavily, some would say exclusively, on outside specialist advice for the solution of all serious technical problems. A third paradox is that architects believe they are offering a service to society but resist the allocation of any great proportion of their professional time to scientific study of its needs. A fourth paradox is that architects place the amassing of personal wealth low on their list of priorities but participate whenever business permits in unconstrained competition with their peers. Finally architects believe passionately in the importance of good design but disagree constantly as to its definition.

If it is true that these attitudes of the professional community have an influence on its relationships with other groups in society,[10] then it may also be true that the positions of those other groups have some impact on the values adopted by the architectural profession itself. The changes we are discussing include changes in the interaction of architects' attitudes and values with those of other professions or other social groups.

Generalization from the observation of these paradoxes in architects' values suggests a broad group of social interactions are at stake. The first two paradoxes relate to the structure of the building industry, to the constraints it places on design and the need for specialization within it. The third and fourth paradoxes are generated by uncertainty over the architect's role —

whether architects should see themselves as artists or as entrepreneurs. The final paradox is concerned with the recognition of quality in building design.

Our review of the content of the official professional journal for the architects of Britain[11] suggests that these broad concerns continued to structure professional debate throughout the period. There were dramatic changes in the ways in which issues which arose in the maintenance of a professional position on each of these concerns were resolved. The framework of relationships between the architectural profession and society at large remained the same, and continuing effort was put into maintaining this overall position. None the less the pressures for change from outside the profession led to many details of the framework being transformed. Indeed it may be that the detailed description of what any profession thinks about itself changes all the time, but that its existence, identity and self-image as an autonomous group have a great capacity to survive.

In the next sections three of the issues discussed by the architectural profession in the 1970s and 1980s are taken as illustrative of this tendency and discussed at greater length. These are: responsibility for the execution of building work, professional ethics and aesthetic judgement.

Construction supervision

In 1968, virtually all building work for which an architect's services were required was undertaken under standard conditions of engagement and a standard form of contract. One of the major features of the contract was that the architect was not only responsible for providing an appropriate design but also responsible for resolving disputes between building owner and building contractor over time delays and cost overruns. Architects' independent judgement in the carrying out of this latter task was thought to be guaranteed by the fact that by practising in partnerships they were financially liable for errors they might make.

Unfortunately architects were also often seen as responsible for errors made by others: those of owners who asked for design features which did not meet the needs of building users, and those of building companies whose materials were unsafe or unstable. As employers of other architects they might also be responsible for errors made by their employees. Theoretically too, employee architects might think themselves responsible for errors made by their employers! These uncertainties had increasing influence on the level of insurance premiums for professional liability throughout the period.

The standard form of contract was also under constant review. In January 1972, for example, revisions were proposed concerning changes in dates of completion, design changes, extensions of time and the role of the architect as a specifier of materials. But in February of the same year, a working party was formed to discuss allowing architects to practise in limited liability companies. In September of that year the Salaried Architects Group pressed the RIBA Council to study responsibility levels of employee architects. In February 1980 new forms of contract were issued which included revised responsibilities for the contractor. By November of that year a form of contract was issued which allowed some of the responsibility for detailed design to be given to the builder. By December the following year a completely different contract was available from a different body, the Association of Consultant Architects, which gave the architect the chance to 'vary [his or her] input'. This was criticized for being attractive in principle, overcoming some liability problems, but avoiding the issue of quality control. In October 1984 another version of the standard form was introduced, and a further report on the costs of liability insurance was released in 1989.

Discussion and debate over the division of responsibilities with others and thus of forms of agreement or contract to be adopted are a perennial part of the architectural profession's culture. There is presumably no best answer on such questions, but always a need to reconsider the present answers and debate new ones.

Professional ethics

The same might be said of the second example of an issue which was widely discussed in the 1970s and 1980s: the architectural profession's code of professional conduct. As has been hinted above, the 1968 ethical position of the code was that professionals must keep themselves clearly independent of commercial pressures. They should not have financial or other interests in building companies, in developer businesses or in component manufacturing firms. Nor should they be seen to behave like business people: they must not advertise their wares and they should not compete with each other on price.

This last factor, the standard level of fee for architects' services, had already come under attack with the publication in July 1968 of a report from the government Prices and Incomes Board. This report had investigated the costs of professional services in construction and stated that although the profession was underpaid, fee income was likely to rise and that the fee scale should be rendered advisory, not mandatory. As indicated above, more commercial forms of business were being discussed as early as April 1968 and the constraint on advertising was subtly altered in November. In some circumstances the name of an architect could now be published but 'only in an unostentatious manner omitting [his or her] address'. By September 1971, a working party was studying limited liability in the UK and overseas.

In 1974 Malcolm MacEwen wrote an article about the crisis in architecture, calling for a reappraisal of the professional function.[12] By the end of 1979 Andrew Rabanek was able to report that analogous code changes introduced in the USA in 1970 had led to better discipline in design and a valuable clarification of responsibilities for time keeping and cost overruns.[13] A complete new code of conduct was issued in January 1981. Advice on the mechanics of practice in limited companies was being offered in December the previous year and the minimum fees, by then mandatory, were finally abolished in March 1982.

In December 1987 a new series of journal articles was initiated which gave practitioners advice on the management of their businesses. Two points can perhaps be usefully made about this almost total volte-face on the role of the professional architect: it took a very long time to achieve, and it was partly the result of continuing strong pressure from the country's largest purchaser of architects' services, the British government. But it must be noted that there still is a code of conduct to which architects should adhere and there still is a fee scale to which they should pay a certain amount of attention. These features of the professional culture still persist and will doubtless continue to be the object of debate and revision.

Aesthetic control

The third example of an issue which was debated throughout the period under discussion is the architects' view on the town planners' power of aesthetic control.

At the beginning of the period under consideration, many official town planners were also qualified architects and could enter into constructive debate with building designers over the aesthetic needs of relevant sites. Problems were beginning to appear with the publication in August 1974 of the Matthews/Skillington Report on design standards in government buildings, some of which do not have to be submitted to planning control. A 1977 article by Siobahn Cantucuzino talked of the horrors which had resulted from twenty years of speculative development[14] and quoted, as examples of what could be achieved, the housing scheme at Alexandra Road in the London Borough of Camden. But other commentators considered this scheme showed the worst of modern design.

In 1979 the minister concerned with the environment, Michael Heseltine, was reported to recommend favouring architects' schemes so that the number of proposals coming to him on appeal could be restricted. In February 1980 the minister stated that delays in granting permission were becoming too great. At the same time he extended the definition of proposals for which no application was needed. In April 1980 a set of good practice details generated in the Greater London Council were published 'pour encourager les autres', but by January of the following year the professional journal could carry an article on the 'frustrated generation'. Good young designers, it seemed, were unable to find sponsors for their innovative ideas. At the same time the older generation was being encouraged to speed up their design process and avoid conflict with an increasingly weak planning system. In October 1983 the journal talked of 'throwing off the sackcloth' and in 1984 the Institute promoted a Festival of Architecture which it was hoped would increase public awareness of new design ideas. A competition for the extension of London's National Gallery led to a compromise design solution being soundly criticized by a member of the Royal Family. In 1985 the great and the good of the architectural profession found themselves unable to defend the design proposed for a new office building and public square in the City against the conservationist establishment. In August 1988 the question of aesthetic control was still high on the agenda of practice issues for the profession. Its aesthetic judgement still had to be defended.

Continuity and change

This review suggests that British architects have contradictory attitudes towards such questions as their position in the building industry, their role in the community and their self-image as artists. In this period of economic and political change, the pressures put upon architects to alter their views led to considerable change in their attitudes towards their responsibility for the building process, their business ethics and the aesthetic quality they expect of their peers. But it may be argued that these transformations have not altered the basic dimensions of the architects' professional identity. It seems that by giving importance to the attacks on their values and by debating the ways in which these may legitimately be altered, the profession can continue to redefine its relationship to other professional and social groups.

The collapse of the modern movement did not mean that architects lost belief in their own aesthetic judgement, it did not mean that they abandoned all pretence at offering an independent service, and it did not mean they had abdicated from all responsibility for the quality of building work. They are still architectural professionals, but they are the architects of 1990 and not the architects of 1970. The world has moved on and they with it.

Notes

1. *The Architect and His Office*, London, RIBA, 1962.

2. Esher, L. *The Broken Wave*, London, Allen Lane, 1981.

3. Watkin, D. *Morality in Architecture*, Oxford, Clarendon, 1978.

4. Dowson, P. Personal communication, 1989.

5. Lyall, S. *The State of British Architecture*, London, Architectural Press, 1980.

6. Gordon, A. 'Architects and resource conservation', in *J. RIBA* Vol. 81, No. 1, 1974, pp. 9–12.

7. Jencks, C. *The Language of Post-Modern Architecture*, London, Academy, 1977.

8. The argument which follows is derived from observations made by Geertz, C. in Chapter 7 of his *Local Knowledge: Further Essays in Interpretive Anthropology*, New York, Basic Books, 1983, on the peculiarities of various other professions including mathematicians and physicists.

9. Cf. ibid, p.72, with Salaman, G. *Community and Occupation*, Cambridge, CUP, 1974, p. 67. The point is repeated by Cuff, D. 'The Ethos and Circumstance of Design', *Journal of Architectural and Planning Research*, vol 6, no. 4, 1989, pp. 305–320.

10. A detailed study in Moulin, R., Dubost, F. Gras, A., Lautman, J., Martinon, J.-P., Schapper, D. *Les Architectes: metamorphose d'une profession libérale*, Paris, Callman-Levy, 1973 demonstrates the variety of roles played.

11. Reference is made in the following pages to reports in the *Journal of the Royal Institute of British Architects*, vols 75–95, (1968–88). At the beginning of this period it was published by the Institute but this responsibility has now been passed to a private company.

12. MacEwan, M. 'Crisis in Architecture', in *J. RIBA*, Vol. 81, No. 4, 1974, pp. 16–45.

13. Rabanek, A. 'The Frontiers of Practice', in *J. RIBA*, Vol. 86, No. 9, 1979, pp. 414–15.

14. Cantacuzino, S. 'Architecture in an Hourglass: the state of the art in Britain today', *J. RIBA*, Vol. 84, No. 2, 1977, p. 49.

Part Two

Architects
Today

3

The profession as a whole

Introduction to the survey

This and the following two chapters consider the work and opinions of architects in private practice in Britain in the early 1990s. Much has been written in the professional press about the changes which have been taking place in the business of architecture through the period which followed the Lords Esher and Llewellyn Davies' 1968 article.[1] Little of this has been founded on solid evidence: it seemed time for a systematic study which would confirm or repudiate myths and provide a sound basis for the evaluation of future possibilities.

The questionnaire survey we undertook obtained the views of 610 architects who were principals in private practice, and is unique. Such detailed information on who architects were, on what they did and on what they thought about their professional situation at a particular time has had to be tailored to a book in which a number of other views of the profession are also made available. The reader is thus able to form a rounded picture of experiential and contextual aspects of practice as well as the operational ones which a questionnaire survey tends to emphasize. In writing these chapters, we have tried to select the information which helps answer the questions about architectural practice posed in the first part of the book and to omit other data which, while of potential interest to some readers, are of less general concern. Appendix II discusses the methodology of the survey and the demographics of the sample. Appendix III provides a copy of the questionnaire.

When this study began, the authors wanted to tell a story. We wanted to tell the story of what an architect does in his or her[2] work and a story about the talents, skills and knowledge one should have to be an architect. A first view might be that all we needed was a look at the curricula of university-level schools of architecture. Surely, one would reasonably postulate, what they teach is well aligned with the knowledge and skills of the professional. As the reader familiar with such curricula will see in this part, this turned out to not necessarily be the case. The popular publicly held image of the architect also might not be accurate. This image is usually characterized as the artistic professional: the architect creates art. Again, this turns out to not be fully accurate.

What do we know about architects and how they work? It seems that the best answer until recently could only be characterized as very little. As discussed elsewhere in this volume, there have been a number of interesting studies of the architect at work. However, no statistically valid studies of the profession as a whole have to our knowledge ever been undertaken in the United Kingdom. The latest approach to this issue was a study published in 1962 by the Royal Institute of British Architects,[3] and it included only a small sample of architects interviewed personally and through a mailed questionnaire. The smallness of its sample left in question whether one could generalize from the sample to the population of architects in the United Kingdom and, given that the information was nearly 30 years old, made clear that much better information was needed. This is especially true when one considers the period of transition that the profession of architecture may now be facing.

The in-depth case studies in this volume give us a very large and, we believe, valuable source of information about a small group of what may be model firms. However, even with that wealth of information, there were still not the valid data from which to construct an overall picture of the profession.

The authors undertook to a large, statistically valid, questionnaire study of principal architects in the United Kingdom to try to fill this critical need for information about how architects spend their time at work, what services the architect provides, how architects see the future of their profession, how well architects feel they were prepared for professional life by university education and training, and how they structure their practices.

General overview of the profession

The questionnaire asked a number of questions which, when assembled, provide an overview of how architects generally see themselves and their profession. The results from these data are presented in this section.

Before beginning, a note on the presentation of these data is warranted. The questionnaire listed a variety of statements. For some of the statements, responding principals were asked to indicate whether they strongly disagreed, disagreed, didn't know, agreed, or strongly agreed with the statement. For purposes of analysis, the strongly agreed and agreed, and the strongly disagreed and disagreed, respectively, were grouped together. Other statements asked responding principals to reply with a scale of very unimportant, unimportant, don't know, important, very important. Similarly, these were regrouped into unimportant, don't know, and important for analysis purposes.[4]

Importance of design and creativity

As indicated elsewhere in this report, the data support the view that design plays a necessary but not sufficient role in the work of an architect. Table 3.1 provides some interesting insight into the actual role of design within the profession. Only three-quarters say they hire new staff based on their talents as designers, clearly indicating that other skills, perhaps in one-quarter of the cases, can be overriding. Yet, nearly 98 per cent of those responding report that the visual aesthetics of the building is important. Thus, this is consistent evidence that, in the view of principals, design, and perhaps style, themselves play critical roles in the practice of architecture but do not encompass the entire professional service.

Creativity appears to be very important to the architect. Eighty-seven per cent are attracted to radical or innovative ideas but only nearly half report that they are not best avoided in favour of the more conventional. Ninety-two per cent report that the chance to be creative is important to them, perhaps distinguishing the architect from other professional fields. Nearly four-fifths report that they are attracted to radical or innovative ideas, yet 30 per cent report that radical or innovative ideas are best avoided. The opportunity to participate in creative work is apparently a motivation for architects. However, this may be in conflict with the desires of the client.

Additional confirmation of the importance of creativity and design is shown by the over 85 per cent who report that, regardless of the fee, architects should uphold their design standards. With nearly 94 per cent believing that the client and architect should reach early agreement on the design aspect of a project, the principals may again be indicating the importance of design: have it agreed upon early so that, at least regarding the design aspects of a project, there are no questions later.

Table 3.1 Importance of design and creativity (per cent (no.))

Statement	Disagree	Don't know	Agree	Mean	Valid cases	Missing cases
Radical or innovative ideas are interesting and thought-provoking [v132]	3.0 (18)	9.6 (58)	87.4 (527)	2.844	603	7
[Radical or innovative ideas] are, in the long run, best avoided in favour of the more conventional [v133]	47.8 (287)	20.5 (123)	31.8 (191)	1.840	601	9
Architects should uphold their design standards, regardless of the fee involved [v135]	7.6 (46)	6.8 (41)	85.6 (516)	2.779	603	7
The architect and the client should reach agreement about the project early in schematic design [v136]	2.6 (16)	3.5 (21)	93.9 (571)	2.913	608	2

Statement	Unimportant	Don't know	Important	Mean	Valid cases	Missing cases
Hiring new staff based on their talent as designers [v49]	15.3 (90)	9.2 (54)	75.5 (443)	2.601	587	23
Visual aesthetics of the building [v61]	0.7 (4)	1.5 (9)	97.9 (592)	2.972	605	5
The chance to be creative [v69]	3.7 (22)	4.2 (25)	92.1 (550)	2.884	597	13

Communication skills are important

The data indicate that principals take an approach to communication that is very open to other points of view. Table 3.2 shows that a variety of communication skills are acknowledged by the principals.

Over 90 per cent agree that preliminary meetings are the most crucial. Nearly three-quarters agree that meetings should be a regular part of communication between architect and client. Over half indicate that they believe in open communications by not wanting separate meetings with the different participants in a project.

Table 3.2 Communication skills are important (per cent (no.))

Statement	Unimportant	Don't know	Important	Mean	Valid cases	Missing cases
Avoidance of conflict with the client, owner or contractor [v66]	6.8 (41)	5.7 (34)	87.5 (525)	2.807	600	10

Statement	Disagree	Don't know	Agree	Mean	Valid cases	Missing cases
The preliminary meetings between the client and the architect are the most crucial ones [v30]	4.7 (28)	3.9 (23)	91.5 (546)	2.868	597	13
Numerous meetings should take place between architects and clients [v32]	23.4 (141)	4.0 (24)	72.6 (438)	2.493	603	7
The architect should meet separately with each participant in a project [v33]	56.0 (332)	7.1 (42)	36.9 (219)	1.809	593	17
Meetings should be used to get client approval only after design issues have been decided by the architect [v34]	65.9 (394)	5.9 (35)	28.3 (169)	1.624	598	12

Over 65 per cent agree that meetings are for genuine exploration of ideas, not for the confirmation of decisions already made. Over 85 per cent report that it is important to avoid conflict with the client, owner or contractor, perhaps indicating the reason for this attitude of openness in decision-making. Yet the complete avoidance of conflict may not be a reasonable expectation.

Architects need to have a special knowledge

As the field of architecture changes and adapts to the changing world, one of its challenges may be to indicate to the society at large and to that society's institutions what the special knowledge of the architect is. As discussed elsewhere, one condition for being a profession is to be perceived as possessing a special knowledge, and knowledge of and ability to design may not be sufficient.

Table 3.3 Architects need to have a special knowledge (per cent (no.))

Statement	Unimportant	Don't know	Important	Mean	Valid cases	Missing cases
Technological innovation [v55]	20.2 (120)	16.0 (95)	63.8 (379)	2.436	594	16
Land usage in design [v56]	12.9 (76)	12.2 (72)	74.8 (440)	2.619	588	22
Application of research to practice [v57]	17.3 (103)	18.6 (111)	64.1 (382)	2.468	596	14
Continuing Education [v58]	7.5 (45)	12.7 (76)	79.8 (478)	2.723	599	11
Project Management [v64]	12.1 (73)	11.8 (71)	76.1 (459)	2.640	603	7
Construction Management [v65]	14.0 (84)	12.0 (72)	74.1 (446)	2.601	602	8
Engineering [v67]	10.3 (61)	17.1 (102)	72.6 (432)	2.624	595	15

Statement	Disagree	Don't know	Agree	Mean	Valid cases	Missing cases
The responsibility falls to the architect to see that the contractor carries out the plans as promised to the client [v119]	12.8 (76)	1.5 (9)	85.7 (509)	2.729	594	16

Whatever the challenges that the profession is facing, the principals report that the special knowledge of the architect is an important aspect of the architect's work, as shown in Table 3.3. Nearly two-thirds report that technological innovation and the application of research to practice is important, indicating a strong role for the special knowledge of the architect. Over four-fifths also believe that the responsibility for the project, should someone else not perform, still falls to the architect, indicating that the architect may need to have knowledge that even falls outside his or her contractual responsibilities.

Nearly four-fifths rate continuing education for the professional as important. Nearly three-quarters report that knowledge of land usage in design, knowledge of project management, and knowledge of construction management and engineering are important. Nearly two-thirds wrote that technological innovation and the application of research to practice are important in architectural practice.

In fact, when push comes to shove in the building process, over 85 per cent of the responding principals report that the final responsibility for the project falls to the architect, presumably regardless of the contractual arrangement. These architects confirm the need for special knowledge in a variety of areas for which they may not have been formally prepared.

Flexibility and personal style are necessary

The principals report that flexibility and personal style are important in the practice of architecture. As shown in Table 3.4, a very significant portion of the principals (nearly half) agree that architects should be prepared to accept a subsidiary role on a project by offering only design services. Yet, nearly 70 per cent do not feel it is important to make clear in publicity materials that they would accept such a role. They are willing to do it but evidently reluctantly.

Table 3.4 Flexibility and personal style are necessary (per cent (no.))

Statement	Disagree	Don't know	Agree	Mean	Valid cases	Missing cases
Architects should always be prepared to offer only a partial service (design only) and allow others to lead the team [v127]	37.9 (228)	13.3 (80)	48.8 (294)	2.110	602	8
It is not always possible to know the tasks a project will involve before beginning work on it [v131]	14.6 (88)	2.3 (14)	83.1 (501)	2.685	603	7
Personal style is important to the success of an architect [v134]	21.3 (128)	11.1 (67)	67.6 (407)	2.463	602	8
Statement	Unimportant	Don't know	Important	Mean	Valid cases	Missing cases
Making clear in publicity material that you will work in a subsidiary design role if necessary [v41]	69.5 (411)	14.7 (87)	15.7 (93)	1.462	591	19

Also, over four-fifths write that projects sometimes change during their execution, presumably indicating the need for flexibility in management of projects, and conflicting with the responses reported earlier that important decisions should be agreed upon with the client early in the process. This inability to know what a project will entail from its inception may reflect a deficit in the knowledge base or it may simply reflect the vagaries of the building process, making firm contractual arrangements and architect and client expectations difficult to achieve. In the continued development of the profession of architecture, these vagaries could prove to be dangerous if they are not clarified.

Lastly, over two-thirds report that their personal style, in their view, is important to the success of an architect. Table 3.5 shows that issues of law, liability and ethics are rated as important by nearly 90 per cent of the principals. These are evidently extremely important concerns.

Table 3.5 Law and ethics (per cent (no.))

Statement	Unimportant	Don't know	Important	Mean	Valid cases	Missing cases
Law and liability [v52]	4.7 (28)	3.7 (22)	89.7 (547)	2.869	597	13
Professional codes of ethics [v53]	6.9 (42)	5.4 (33)	87.7 (533)	2.808	608	2

Anyone who has attended a school of architecture knows about the tradition of the 'all-nighter'. As shown in Table 3.6, the principals report that such work schedules continue into practice.

Table 3.6 Work habits (per cent (no.))

Statement	Disagree	Don't know	Agree	Mean	Valid cases	Missing cases
I frequently work in the evenings and on weekends [v123]	13.9 (84)	2.2 (13)	83.9 (506)	2.700	603	7
It is difficult to find more than 30 minutes of uninterrupted time in a working day [v124]	28.2 (169)	6.7 (40)	65.2 (391)	2.370	600	10
Setting a workable time schedule for projects is a top priority for my firm [v125]	19.0 (114)	13.5 (81)	67.5(405)	2.485	600	10

Over four-fifths report that they often work evenings and weekends. Over 65 per cent report that it is difficult to find even small periods of uninterrupted time. About the same percentage report that setting workable schedules is a top priority. They appear to maintain the approaches to work that they learn in school but recognize that a problem persists. Surely this creates a significant problem in the management of the architect's office.

Marketing expertise is necessary

The marketing of architectural services is little covered in the education of an architect. The principals indicate, as shown in Table 3.7, that there is little consistency in approaches to the subject and that most rely on that most mysterious and ill-defined of marketing approaches, personal contact.

Table 3.7 Marketing expertise is necessary (per cent (no.))

Statement	Unimportant	Don't know	Important	Mean	Valid cases	Missing cases
Talking regularly with land surveyors, engineers, contractors and others to get leads on clients [v42]	36.1 (216)	11.0 (66)	52.8 (316)	2.167	598	12
Keeping up with real estate news and local planning reports to learn about possible new clients [v43]	39.5 (236)	10.7 (64)	49.7 (297)	2.102	597	13
Visiting and leaving information with potential clients to establish a rapport before they need an architect [v44]	39.1 (232)	13.5 (80)	47.5 (282)	2.084	594	16
Calling clients informally about some item outside of architecture of interest to the client [v45]	41.9 (250)	15.3 (91)	42.8 (255)	2.008	596	14
Inviting the client out socially [v46]	45.3 (268)	11.8 (70)	42.8 (253)	1.975	591	19
Globalization of the profession [v51]	44.1 (241)	38.4 (210)	17.6 (96)	1.735	547	63
Statement	Disagree	Don't know	Agree	Mean	Valid cases	Missing cases
Private connections and contacts often lead our firm to clients [v122]	9.0 (54)	4.2 (25)	86.8 (519)	2.778	598	12

Only about half or fewer of the principals report that talking regularly with land surveyors, engineers or contractors, keeping up with real estate news or local planning reports, personally visiting potential future clients; meeting potential clients informally, or inviting clients out socially, leads to clients. Yet, nearly 87 per cent report that private connections and contacts often lead to clients for the firm.

It is of interest to note that only 17.6 per cent of the principal architects reported that the globalization of the profession was important. With the increasing integration of the European Union, which of course includes the United Kingdom, one would think that both the expanded competition and the expanded opportunity would be on the minds of architects. Perhaps, given the generally small firm size, as discussed elsewhere, this expansion of the potential market area is insignificant to most architects.

Statistical cross-tabulation of the responses to the six statements about methods of developing clients reveals that firms that are likely to use one method other than personal contact are also likely to use all the approaches in addition to personal contact ($p < 0.01$). It would appear from these statistically significant differences that some firms have marketing plans and approaches and others simply do not. Interestingly, this would mean that the majority of firms (if only a slight majority) do not have any planned approach to marketing their services. In a context of weak demand and when there is an increasing emphasis on business approaches, this is very surprising. It is more likely that the majority of firms are simply not organized to undertake directed marketing efforts.

There was not a statistically significant difference between responding principals who reported using contacts for marketing and responding principals who reported using any of the other approaches so far as their view on the globalization of the profession is concerned. Preparation for a global market for architects does not seem to have any uniform pattern.

Notes

1. See p. vii and note 1 to the Preface.

2. Only 4 per cent of the respondents were women. On 29 January 1995, the *Independent on Sunday* published a report of a survey which showed that only 9 per cent of Britains's 30 000 architects to be female. It stated that 'the RIBA has yet to adopt an equal opportunities or sexual discrimination policy'.

3. *The Architect and His Office*, London, RIBA, 1962.

4. The codes [v132] etc. refer to the statement numbers in the questionnaire in Appendix II. In some tables there is a column headed 'Mean'. This refers to the arithmetic average of responses marked [1] [2] [3] etc. in the questionnaire.

4

Organization and management

Organizing the work

Several of the questions on the questionnaire address various aspects of the organization of the work of the architectural professional. These questions ask responding principals to report what portion of the average work week is spent on a variety of activities, their perceived adequacy of their formal education, as well as other items. Also, analysis indicates that there are differences in how the work is organized and in the principals' views of the profession depending on the size of the firm, whether the firm specializes in a particular building type (and which specialization that is) and whether the practice is a multi-disciplinary firm.

Before reporting these results, a few definitions are in order. Throughout this report reference is made to the 1962 RIBA study *The Architect and His Office*.[1] Although this covered a smaller number of practices it is the most recent comparable study. Parallels with some of its results are reported here to show how much has changed over the 30-year period.

The RIBA study divided firms into size groupings in two different ways. One method was a three-way split: small firms of 1 to 10 architects; medium firms of 11 to 30 architects; and large firms of 31 or more architects. The other method used by the RIBA study was a five-category split: 1–5, 6–10, 11–30, 31–50 and 51 or more. Where possible this chapter duplicates these groupings in order to provide comparable data. (The questionnaire itself, shown in Appendix II, simply asked responding principals to report how many architects, planners, quantity surveyors, designers, landscape architects, etc. were employed by the firm. These responses were recoded during analysis into the above noted groupings.)

For the purposes of this study, a practice is considered multi-disciplinary if the principal responding reported that at least one non-architect professional staff member was employed by the firm. The sections of this chapter discuss each of these subjects separately.

How principals spend their work time

How principals spend their time at work can be interpreted as a strong indicator of the knowledge base that architects use; Table 4.1 analyses this.

Table 4.1 How principals spend their time at work: a comparison of 1962 and 1991

				Total architectural staff				
Firms	1–10	**1–10**	11–30	**11–30**	31+	**31+**	All	**All**
Year of study	1962	1991	1962	1991	1962	1991	1962	1991
Number of responses	13	540	9	50	8	15	30	605
Per cent of year's total	43	89	30	8	27	3	100	100
Activity category (%)		(* indicates management activity)						
*Travelling[†]	3.2	3.3	5.5	5.5	7.0	6.9	¶	3.4
Drawing	32.1	28.8	29.1	15.9	12.7	8.8		33.3
*Meetings and discussions	25.0	24.9	22.8	41.2	50.0	41.0		21.0
Site supervision	2.5	8.8	5.0	3.0	0	2.7		7.5
*Correspondence, preparation of reports	18.5	16.9	19.2	15.6	13.5	17.2		17.7
*Staff supervision	13.7	2.8	8.2	4.4	7.2	5.4		2.9
*Technical references	0.2	6.4	1.1	5.0	1.0	7.4		6.1
*Filing, general office activities, etc.[‡]	3.9	8.3	8.4	9.4	7.0	10.6		8.1
Totals[§]	99.1	100.2	99.3	100.0	98.4	100.0		100.6
*Management total	64.5	62.6	65.2	81.1	85.7	88.5		59.2
% change in management total		−2.9		+24.4		+3.3		

† In the 1991 data it was necessary to use different categories than were used in the 1962 study. First of all we wanted to obtain more specific information. Categories such as drawing and travelling were seen to be unnecessarily vague. Also, the very nature of modern office work has dramatically changed since 1962. For example, the activity of reading technical references has, in some cases, changed to the activity of conducting research. The computer in the workplace may have produced many other changes. However, in order to compare the 1991 data with the 1962 data, the categories of the 1991 data were regrouped to match the 1962 data.

These groupings are as follows:

1962 activity category	1991 activity category
Travelling	Recruiting clients in person
Drawing	Building design
	Production drawings
Meetings and discussions	Co-ordinating consultants
	Meeting with clients
	Meeting with project managers
	Meeting with marketing specialists
	Negotiating new work
Site supervision	Site supervision
Correspondence,	Agreement writing
preparation of reports	Specification writing
	Publicizing work
	Recruiting clients on the telephone
	Other approaches to recruiting clients
	Establishing office fee structure
Staff supervision	Staffing office
Technical references	Construction budgeting
	Work estimation
Filing, general office activities, etc.	Developing office procedures
	Office finance
	Other management activities
	Other activities

‡   In the 1962 RIBA study, this category was reported as three categories: filing, etc.; general office activities; and absence, etc. The 1991 study did not ask about absence since the meaning of absence in the 1962 study was so unclear. Therefore, to compare the 1962 figures with the 1991 results, the absence figure needed to be deleted and the remaining 1962 figures distributed in the same proportions. Also, the categories of filing and general office activities from the 1962 study are combined as reported here.

§   Totals other than 100 per cent are due to rounding errors.

¶   These figures were not reported in the 1962 RIBA study.

At the end of the day, no matter how pervasive the public image of the architect as the gentleman artist, the ways that an architect spends his or her time strongly indicate the skills required for the successful conduct of practice. Also, the work of principals is largely self-directed. Unlike lower-ranking assistants who may be assigned tasks by others in the firm, principals have both greater latitude and greater responsibility in deciding how the time of an architect should be allocated. Table 4.1 shows how principals, controlling for firm size, spent their time in the 1962 RIBA study[2] and how, today, they report they are spending their time. These data demonstrate some very interesting changes in how principals in architecture firms are spending their time. First it must be noted, however unfortunate it is, that the small sample size of the 1962 study may mean that its results are not representative of the profession at large. It may not have been intended to be representative, in fact. Yet, it provides the only comprehensive points of comparison.

Perhaps the most striking difference shown in Table 4.1 is the amount of time spent on management activities. While, as reported elsewhere, building design is reported to be of critical importance to the responding principals, in large firms, for example, nearly 90 per cent of a principal's time is spent on management activities. In the medium firms this figure jumps from 65.2 per cent to 81.1 per cent indicating a significant increase in management activities. For mid-size firms, time spent on drawing drops from 29.1 per cent to 15.9 per cent while time spent on meetings and discussions leaps from 22.8 per cent to 41.2 per cent.

One should not be too quick, however, in deciphering these figures. Yes, it is clear that a significant portion of an architect's time is spent on management activities for which the architect is ill-prepared in university. But, it does not necessarily follow that university curricula should proportionately match how the average architect's time is spent. It does however indicate that increases in the preparations for these management activities are strongly warranted. Of course, one could argue that the architect could drop all management activities and be only the building stylist. Yet, this would not only eliminate the reasons for legal sanction for the architecture profession (see below) but would also drastically diminish the services an architect provides and, therefore, the perceived need for an architect at all – not a particularly joyful prospect.

Second, the 1991 study covered a more representative sample of the profession and included a substantial number of small-firm practitioners. These were not fully covered in the earlier study.

Activities reported as involving travel remained constant over the 30-year period for all sizes of firm. Since the principal in the larger firm is more able or likely to specialize in his or her activities, it is not a surprise that such principals spend about twice as much time travelling as principals in small firms.

For all three categories of firm size, the amount of time spent drawing by principals dropped over the 30-year period. For the medium firm, this drop was nearly half. This again would lend credence to the growing importance of management activities in the practice of architecture.

The amount of time spent on site supervision went up significantly for principals in small and large firms but went down for principals in medium firms. One can only conjecture why this occurred. Also, in all three cases, it is still a small proportion. Yet, it should be noted that despite the commonly held notion that architects are removing themselves from actual construction activities, some principals still provide these services.

The amount of time spent on staff supervision went down in all three firm size categories. Perhaps this decrease reflects the widespread introduction of computers in the workplace and concomitant reductions in the numbers of technicians and support staff. Also, this may be an indication of limited employment opportunities for new university-level graduates.

There is a dramatic jump between the two surveys in the amount of time spent by principals on technical references. While the total percentage is small, the complexity of the building process has increased over the 30-year period and subsequently the amount of time necessary to keep abreast of new developments has increased. The demands of legal liability and insurance requirements may have also influenced an increase in the focus on technical references. Yet, the conditions for legal sanction would appear to require such an increase in focus. Additionally, the amount of time spent on general office activities has increased (which probably included project documentation, record keeping and related insurance requirements).

The public image of the architect may be that he spends his time in drawing and creating (what is generally called design), an image that is reinforced by the traditional university curriculum. However, these data clearly report that: the architect is spending less time on building design and drawing than he or she did 30 years ago; and time spent on building design ranges from a low of 8.8 per cent to a high of 28.8 per cent in the smaller firms. This would indicate that the average architect, in any size firm, spends a maximum of about 12 hours per week on design-related activities. This is a far cry from that traditional public image of the architect.

Perceived adequacy of education

Building design itself fills a significant but less than majority portion of a principal's time. Given that training in design most commonly forms the overwhelming majority of the education of an architect, one would hypothesize that architects might report that their education did not adequately prepare them for professional practice. To gain some indication of their views on this, the architects receiving the questionnaire were asked to report whether they 'received adequate training in' each of the subject areas shown in Table 4.2.

Before reviewing Table 4.2, it is worth noting that this question asked about the *adequacy* of a respondent's education. For example, a little needed subject could have been very sparsely covered yet still be reported as completely adequately covered. A low rating does not indicate that the subject is highly important to an architect in practice, only that the subject received less coverage than it should have received.

Table 4.2 Principals agreeing they received adequate training in given subject areas (per cent* (no.))

Schematic design	85.4 (496)	Project management	21.3 (118)
History of architecture	87.3 (507)	Client relations	15.2 (85)
Building technology	80.6 (469)	Office management	15.2 (85)
Structural/mechanical design	74.9 (438)	Budget management	14.2 (78)
Urban design/planning	58.8 (342)	Computer-aided design	12.0 (64)
Interior design	51.0 (292)	Real estate development	11.9 (65)
Specifications and codes	50.9 (295)	Accounting	10.5 (58)
Production	49.8 (271)	Facility management	10.3 (54)
Brief preparation	48.5 (273)	Marketing	8.5 (47)
Research	41.4 (229)	Computerization	8.1 (43)
Human behaviour	40.6 (227)		
Communication	31.8 (180)		
Construction management	29.7 (166)		

* The percentage represents the portion of the total responding to that individual question who agree with the statement. As is usual on questionnaires, some responding principals skip individual questions. Therefore equal percentages for different questions can represent a different absolute number of responding principals.

From Table 4.2 one quickly sees a pattern which confirms the hypothesis that an architect's education may not match what an architect needs for professional practice. One readily sees that building schematic design, history of architecture and building technology are covered to the satisfaction of a very high percentage of the responding principals. About three-quarters and three-fifths report that structural/mechanical design and urban design/planning, respectively, are adequately covered.

However, after those top five categories, the percentage agreeing that they received adequate training in a long list of topics which, as described earlier, they also feel are an integral part of the work of a practising architect, drops off very quickly. The management areas which occupy the majority of the practising architects' time (see Table 4.1) all receive notably low ratings. Marketing, accounting, real estate development, budget management, office management, client relations, project management, construction management and communication *all* receive very low ratings for the adequacy of the university-level training received by the principals. These management-related areas show significant deficits in the education and training of an architect.

Professional legitimacy

What makes a profession legitimate in a modern society? How does it maintain the market for its services? How does it maintain legal sanction and market protection for the services it provides? In the past few years (especially with the recent proposed deregulation of architects in the United Kingdom) these are questions of increasing importance. It can be said that there are three primary conditions which legitimate the existence of a profession and separate that profession from simply being members of a shared occupational class. All three of these conditions are necessary (if not necessarily sufficient) for professional status. These three conditions are:

- *Special knowledge:* Members of this occupational class, by either training, education or experience, have obtained a knowledge base which is not generally known to others. They have also obtained the skills necessary to apply that knowledge base in the service of their clients and the society at large. Additionally, there must be special knowledge not only in the eyes of the service providers. The service recipients, the public at large, must also perceive that (1) a special knowledge exists; (2) they need that special knowledge; and (3) the special knowledge is not generally available from people who are not members of this occupational class.

- *Protection of the public interest including public health, public safety and public welfare:* The special knowledge of this group is needed by the society and its members for the protection of the public and the members of the society are willing to pay for the application of the special knowledge. If those who did not have the knowledge base were to attempt to provide this service, harm to the recipient of the service would be likely.

- *Legal sanction*: Individuals who have this special knowledge are in this way identified to the public as those who are capable of providing the knowledge and skills. This may take the form of a licence, registration, or other governmental (or private) unit assuring that those who purport to provide the service are capable of doing so.

One must admit, these are very general conditions. Physicians, attorneys, and accountants meet these three conditions. But, do architects?

Over the past 40 or so years, the public perceptions of the architect may have been increasingly focused on the artistic or sculptural aspects of the work of the architect. The technical aspects of the architect's tasks have not been promoted in the public's view as the reason one needs an architect. It has been said that it is the design that sells a building. The technical aspects of the

45

building process have begun to be more frequently offered by non-architects: engineers, project managers, facility managers, etc. This may create incontestable dangers for architecture as a profession. No Western society licenses (or provides legal sanction in any other way) for its artists. Nor should it. But, if the architect is viewed primarily as an artist, professional legitimacy may be facing serious challenge. This section discusses not only the areas of special knowledge which architects believe they provide but also how, in the responding principals' views, the special knowledge of an architect should in the future continue to extend far beyond the artistic aspects of the building process.

Future educational needs

What will architects need to know in the future? The questionnaire asked the principals to indicate in which subject areas an architect should receive training in the future. If, for example, they report that an architect needs training in an area, it could indicate that this is an activity which they regularly pursue as part of their practice. Additionally, the responses to this question provide some of the responding principals' speculations about the future of the architect. An overview of these results is shown in Table 4.3.

Table 4.3 Principals agreeing an architect needs to receive training in given subject areas (per cent* (no.))

Building technology	97.0 (557)	Project management	79.1 (433)
Schematic design	95.0 (535)	Construction management	78.0 (447)
Brief preparation	93.2 (534)	Production	77.9 (429)
Specifications and codes	91.6 (522)	Interior design	77.0 (438)
Communication	90.6 (523)	Human behaviour	74.0 (416)
History of architecture	86.0 (491)	Computerization	69.0 (389)
Computer-aided design	85.5 (490)	Marketing	63.3 (357)
Structural and mechanical design	84.1 (481)	Research	61.8 (346)
Urban design and planning	80.8 (463)	Accounting	57.1 (319)
Office management	81.9. (461)	Real estate development	46.3 (259)
Budget management	81.2 (458)	Facility management	43.8 (235)
Client relations	79.7 (456)		

*   See note to Table 4.2.

Table 4.3 gives a very clear indication about the perceived training needs of architects. It is obvious that architects need training, as is traditionally done in schools, in building technology, schematic design, brief preparation, and specification and codes. This is no surprise. Perhaps it is important to note that building technology ranked slightly above schematic design. Some schools of architecture over the past twenty years have reduced the technological aspects of their curricula. It has been said that students come to architecture school to enter a profession that permits them to be an artist, shunning the technological subjects. If such motivations and aspirations on the part of students is true, these principals would emphatically not agree. In fact, in view of the supreme ranking given by the principals to building technology, these students might be well warned to seek employment elsewhere. Lastly, it is possible that the principals are giving a warning that unless architecture students are well prepared in building technologies, we may see aspects of the building process now performed by architects being performed by non-architects, accelerating the diminution of the role of the architect in the building process.

Notwithstanding, we quickly see areas in which the principals report an architect needs training but which are rarely a significant portion of the architecture curriculum. While schools usually address building technology and history of architecture in depth, the other subjects receive minimal or simply no coverage. It would appear that computer-aided design, office management, budget management, construction management, human behaviour, marketing, research and accounting as well as many other areas of expertise are needed by architects, reflecting the activities that the responding principals pursue. One could argue that many of these subject areas are covered in the training of an architect in the traditional year-out programme for architecture students. However, this must be seen as weak coverage at best: the subjects are likely to be learned in a haphazard manner under the tutelage of someone who also learned them in a haphazard manner. Given that the responding principals report a very strong need for expertise in these areas, such a haphazard approach hardly seems appropriate.

As discussed earlier, these data provide additional confirmation of the importance of management skills both in the profession of architecture and in educating practitioners.

Personal satisfaction

In a variety of ways, architects derive significant personal satisfaction from their work. They see themselves as providing a positive influence in society, would generally not leave architecture and believe they are serving their clients. This section reports on attitudes that may lead to this general personal satisfaction.

Architects serve public good

As shown in Table 4.4, over three-quarters of those responding to this question wrote that they believe in the importance of the education of the client and the public. Nearly 70 per cent reported that it is important to promote architectural thinking.

Table 4.4 Architects serve public good (per cent (no.))

Statement	Unimportant	Don't know	Important	Mean	Valid cases	Missing cases
Education of the client and the public [v62]	8.6 (52)	13.1 (79)	78.3 (474)	2.698	605	5
Promotion of architectural thinking [v63]	12.3 (74)	17.9 (108)	69.8 (420)	2.575	602	8
Statement	Disagree	Don't know	Agree	Mean	Valid cases	Missing cases
Most architects adhere to strict professional codes of ethics [v126]	18.6 (111)	23.3 (139)	58.1 (347)	2.395	597	13

Nearly 60 per cent believe that most architects adhere to strict professional codes of ethics. This indicates a strong motivation to serve the public good.

Architects serve clients' needs

As continued in Table 4.5, nearly 95 per cent of architects believe direct contact with the building user is important. As increasing numbers of buildings are built speculatively, without known users, this response becomes increasingly interesting. Whatever the building process, the desire for direct contact with building users is clearly a desire to serve the needs of those users. Over 80 per cent report that they want staff with a wide variety of capabilities. While, as discussed above, design talent is important, here is further indication that design skills, alone, are not enough. Just under half wrote that the cultural diversity of clients is not an important factor. This is particularly perplexing given the multicultural society now developing in the United Kingdom.

The United Kingdom has seen growing ethnic diversity in its population like many other Western industrialized countries. About 35 per cent report that this diversity is important, perhaps reflecting a growing recognition of this important factor in the populations the architect serves.

Table 4.5 Architects serve clients' needs (per cent (no.))

Statement	Unimportant	Don't know	Important	Mean	Valid cases	Missing cases
Direct contact between the user and the architect during the design process [v47]	2.9 (17)	2.3 (14)	94.8 (565)	2.919	596	14
Hiring new staff with a wide variety of capabilities [v50]	10.2 (60)	6.5 (38)	83.3 (489)	2.731	587	23
Cultural diversity of clients [v54]	45.5 (270)	19.9 (118)	34.7 (206)	1.892	594	16
Habitability of the completed structure [v59]	1.7 (10)	2.7 (16)	95.6 (570)	2.940	596	14
Client satisfaction [v60]	0 (0)	0.3 (2)	99.7 (603)	2.997	605	5
Serving people's needs [v68]	1.0 (6)	4.8 (29)	94.2 (566)	2.932	601	9

Statement	Disagree	Don't know	Agree	Mean	Valid cases	Missing cases
The desire to produce excellent projects is often overshadowed by client limitations/demands [v120]	13.8 (83)	5.5 (33)	80.8 (487)	2.670	603	7
Architects always place public interest above client satisfaction [v121]	80.9 (482)	12.6 (75)	6.5 (39)	1.257	596	14

None the less, over 95 per cent rate the habitability of the completed structure as important; nearly 100 per cent rate client satisfaction as important; nearly 95 per cent rate serving people's needs as important: all indicating an extremely strong desire to provide a service to the client and user populations. In fact, just over 80 per cent report that architects do not even place the public interest above client satisfaction, again indicating the strong intention to serve the client. (Of course, it also makes one wonder whether important aspects of the public interest are thus sacrificed.) The principals also report, however, that the client can sometimes limit the excellence a project is able to achieve. Thus, there appears to be a conflict between the importance of serving the client and the perception that clients limit excellence. Perhaps, also, the realization that the client and the building user are very often different people or groups is a relatively new phenomenon and not yet well accommodated within the architect's design process.

Overall report

Overall, the principals report a picture that suggests they are happy with their chosen line of work and, if given the choice, they would do it again. This is surely strong confirmation of the personal satisfaction they receive from being architects.

Table 4.6 Overall report about the profession (per cent (no.))

Statement	Disagree	Don't know	Agree	Mean	Valid cases	Missing cases
Architecture today is more a business enterprise than a profession [v139]	30.2 (183)	12.9 (78)	56.9 (344)	2.266	605	5
Firms which focus entirely on design have become obsolete [v140]	51.3 (307)	27.4 (164)	21.2 (127)	1.699	598	12
The increased demand for architectural services over the last decade gave me greater choice in projects [v141]	43.8 (265)	18.2 (110)	38.0 (230)	1.942	605	5
Greater competition in the architectural field has forced me to explore new approaches to architectural practice [v142]	34.4 (208)	11.4 (69)	54.1 (327)	2.197	604	6
Computers will be vital to architecture in the future [v143]	11.1 (67)	14.5 (88)	74.4 (450)	2.633	605	5
Today's architectural firms must offer more comprehensive services than were necessary in the past [v144]	12.6 (76)	10.6 (64)	76.9 (465)	2.643	605	5
If starting over, I would become an architect again [v145]	14.9 (90)	15.6 (94)	69.5 (420)	2.546	604	6
I would be willing to work outside of architecture for more money [v146]	56.0 (332)	20.7 (123)	23.3 (138)	1.673	593	17
I could foresee myself moving completely out of the architecture field for my livelihood, for reasons other than illness or retirement [v147]	58.6 (355)	15.0 (91)	26.4 (160)	1.678	606	4
I am very satisfied with my career [v148]	13.0 (79)	13.0 (79)	74.1 (451)	2.611	609	1

Just over half wrote that architecture today is more of a business than a profession. Just over half feel there is still a role for firms that focus only on design. About 40 per cent report that the boom period of the 1980s gave them greater choice in projects. However, just over half report that increased competition led them to explore new approaches, an innovativeness which, as reported above, they said they desire.

Nearly three-quarters of the principals recognize that computers will be vital in architecture in the future. Just over three-quarters believe that architectural firms must offer more comprehensive services in the future, perhaps again indicating the limitations of the design only-firm.

But, most pleasing are the responses about life as an architect. Nearly 70 per cent, if starting over, would become an architect again. Given the limited financial rewards, the other satisfactions of the profession must be significant. Over half report that even if they were given the opportunity to work outside of architecture for more money, they would turn down the offer. It is often noted that the education and training of an architect are broad enough to serve many careers. However, only about one-quarter of the principals could even see themselves ever moving out of architecture to earn their livelihood. Just under three-quarters state that they are very satisfied with their career.

Apparently, whatever their other trials, frustrations and limitations, architects generally appear well satisfied with their professional lives.

Notes

1. *The Architect and His Office*, London, RIBA, 1962.
2. Ibid., p. 218.

5

Alternatives and choices

Differences by firm size

One might wonder whether (or, perhaps, expect that) principals of larger firms would report different opinions from those of smaller firms. For example, principals from larger firms might be more likely to recommend specialization in the university-level education of architects. They might be more likely to recommend that architects learn greater management skills. Or, they might be less likely to recommend that entire firms specialize (e.g. the design-only practice). There might be other differences. For this reason, the responses were examined based on firm size.

As discussed in Appendix II, there are a variety of ways to measure firm size. However, for our purposes here, the number of qualified architects in the firm was used. A three-way split was created, which parallels the 1962 RIBA study.[1] Firms were categorized into those with 10 or fewer architects (referred to below as 'small' firms); those with 11 to 30 architects ('midsize' firms); and those with over 30 architects ('large' firms). Then, the responses to the other questions in the questionnaire were compared. Those results are reported here. (All differences reported here are statistically significant as defined in Appendix II.)

Specialization of the firm

For purposes of analysis, a firm was considered to have developed a specialization in a particular area if it spends more than 25 per cent of its time conducting work in that area of expertise. While there are statistically significant differences, they are not exactly what one might expect, as seen in Table 5.1. Smaller firms were more likely to have developed specializations in housing projects. Larger firms were more likely to have developed specializations in commercial or industrial projects. Yet, when it came to doing work for the individual client, small firms were the most likely, large firms the next likely, and midsize firms the least likely. Thus, it would appear that firms are definitely developing building type specialities and that the choice of speciality does have some connection with the size of the firm. Other areas of specialization (institutional and public buildings and urban design) showed no statistically significant difference among firm sizes.

Table 5.1 Specialization of the firms:* differences by size of firm (per cent (no.))

	Firm size		
	<11 (Small)	11–30 (Midsize)	31+ (Large)
Specialization area:			
Housing	46.6 (199)	14.6 (6)	11.1 (1)
Commercial/industrial	46.2 (204)	58.0 (29)	73.3 (11)
Individual clients	40.3 (188)	19.5 (8)	25.0 (3)

*   Specialization is defined as a firm reporting that it spends over 25 per cent of time on the named area.

The work of the principal

There were some differences in the ways that principals spend their time at work depending on the size of the principal's firm, as shown in Table 5.2.

Table 5.2 Time spent by the principal: differences by size of firm * (per cent (no.))

	Firm size		
	<11 (Small)	11–30 (Midsize)	31+ (Large)
Activity:			
Building design	37.6 (205)	14.0 (7)	13.3 (2)
Building production drawings	30.3 (165)	4.0 (2)	0.0 (0)
Construction site supervision	16.1 (88)	2.0 (1)	6.7 (1)
Co-ordinating consultants and staff	11.4 (62)	34.0 (17)	33.3 (5)
Writing accurate specifications	3.9 (21)	0.0 (0)	0.0 (0)
Recruiting clients in person	2.4 (13)	2.0 (1)	20.0 (3)
Meeting with project managers	2.0 (11)	4.0 (2)	0.0 (0)
Recruiting clients by mail	1.7 (9)	4.0 (2)	0.0 (0)
Staffing office	1.4 (7)	0.0 (0)	0.0 (0)
Developing office procedures	0.4 (2)	4.0 (2)	0.0 (0)

*   The figures represent the principals, by firm size, reporting that they spend 20 per cent or more of their work week on the listed activity: see Appendix III for exact wording.

The larger the firm, the more likely the principal is to spend a generous amount of his or her time publicizing completed work; working with marketing or public relations specialists; co-ordinating consultants and staff; or recruiting clients in person. Yet, the principal in the smaller firm is more likely to spend a substantial part of his or her time on building design; on building production drawings; on construction site supervision; or writing specifications. The principal in the midsize firm is more likely to spend his or her day meeting with project managers or recruiting clients by mail.

What does this mean? It would appear that the principal of the large firm is more likely to be involved in the managerial aspects of architectural practice while the principal of the smaller firm is more likely to stay involved in building design and construction activities. This is not an unlikely consequence. Larger firms (in any field) often require the specialization of those in the higher strata of the organization.

One should also note that, with a few exceptions, the actual numbers and percentages of those who reported spending 20 per cent or more of their time on a particular activity are quite few. Additionally, the other categories of work listed on the questionnaire showed no statistically significant difference among firm sizes. These categories of work included meeting with clients; getting agreements in writing; monitoring construction budgets; estimating work remaining on a project; recruiting clients by telephone; negotiating/contracting new work; staffing the office; establishing the office fee structure; and managing office finances. Overall, one can conclude that there are differences in the work of the principal depending on the size of that principal's firm, but that those differences are not as strong as one might expect. A response may, of course, reflect rather the work of the individual principal answering the questionnaire than the work of that individual's firm generally.

Organizing one's work

The principals were asked a number of questions about how one best organizes the work of the architect. One would expect that larger firms organize themselves differently from smaller firms and that these differences would appear in the responses. This area is considered in Table 5.3.

These approaches to and attitudes about work in the architect's office do show some differences among different size firms. The smaller the firm the more likely the principal is to report that the preliminary meetings between the client and the architect are the most crucial ones and that private connections and contacts often lead the firm to clients. In addition, principals in smaller firms are more likely to report the following are important: avoidance of conflict with the client, owner or contractor; and construction management.

The larger the firm the more likely the principal is to report that the following factors are important: hiring new staff based on their talents as designers; keeping up with real estate news and local planning reports to learn about possible new clients; visiting and leaving information with potential clients to establish a rapport before they need an architect; calling clients informally on a subject outside of architecture of interest to the client; inviting the client out socially; the cultural diversity of clients; the globalization of the profession; and joining client trade associations or business and service organizations to establish rapport with clients.

Table 5.3 Organizing one's work: differences by size of firm * (per cent (no.))

	Firm size		
	<11 (Small)	11–30 (Midsize)	31+ (Large)
Agreeing with statement:			
Preliminary meetings are most crucial	92.3 (492)	84.0 (42)	85.7 (12)
Private contacts lead to clients	88.2 (473)	79.2 (38)	57.1 (8)
Reporting as important:			
Avoidance of conflict	88.2 (473)	88.0 (44)	57.1 (8)
Construction management	77.1 (415)	50.0 (25)	42.9 (6)
Hiring talented staff	72.8 (380)	96.0 (48)	100.0 (15)
Keeping up with real estate news	48.2 (257)	64.0 (32)	57.1 (8)
Visiting potential clients	44.3 (235)	78.0 (39)	57.1 (8)
Calling clients informally	41.4 (220)	48.0 (24)	78.6 (11)
Meeting clients socially	38.1 (201)	80.0 (40)	85.7 (12)
Cultural diversity of clients	32.5 (172)	46.0 (23)	73.3 (11)
Globalization of the profession	16.1 (78)	24.0 (12)	42.9 (6)
Joining client organizations	15.9 (85)	34.0 (17)	57.1 (8)
Working in a subsidiary design role	12.7 (67)	46.9 (23)	24.4 (3)

*   The figures represent the principals, by firm size, agreeing with or reporting as important a particular approach to work organization: see Appendix III for exact wording.

Lastly, it appears that the midsize office is the most likely to be willing to make clear in publicity material that it will work in a subsidiary design role if necessary.

Overall, these data build a clear pattern. The larger the firm, the more likely it is to be recognizing and responding to the changes in demographic diversity occurring in the United Kingdom and the economic change occurring in the European Community and elsewhere. Additionally, the larger firms are more likely to have developed multiple strategies to marketing their services and obtaining work. Smaller firms depend more heavily on private contacts while the large firms have developed multi-faceted methods of making their services known to potential clients. One must ask if this approach to marketing is, perhaps, how these became the larger firms. The smaller firms, unable to distribute the work based on specialized expertise within the office, are, perhaps, unable to mount effective marketing campaigns. Lastly, it is interesting that the midsize firm is most willing to work in a design-only subsidiary role. Perhaps, when compared with large offices, this is because the large offices desire all aspects of a project since they have developed a wide variety of capabilities afforded by their size and the concomitant ability for staff to develop specialized skills. However, why the small firms report an unwillingness to work in a design-only subsidiary role remains unclear. It would appear to be a natural source of work for small firms.

The location of meetings appears to have a statistically significant difference among the sizes of firms, as illustrated in Table 5.4. The larger firms are much more likely to conduct meetings in the office than other size firms. The small firms are much more likely to conduct meetings at the project site. This reflects and is in concurrence with the earlier reported findings that the principal in the small office is more likely to be directly involved with construction. It also may reflect the inability of the small office to permit specialization among its professional staff.

Table 5.4 Location of meetings: * differences by size of firm (per cent (no.))

	Firm size		
	<11 (Small)	11–30 (Midsize)	31+ (Large)
Meeting location:			
Office	40.1 (201)	76.0 (38)	66.7 (10)
Project site	48.8 (246)	19.0 (8)	00.0 (0)

*   Location of meetings more than 25 per cent of the time.

What training do architects need?

The size of the firm of the responding principal also showed some differences in the areas in which that principal thought an architect should be trained. This is considered in Table 5.5. Except for real estate development, training in which is favoured by principals from midsize and large firms, the remainder of these differences by firm size show a puzzling pattern. The small and large firms' principals are more likely to believe an architect needs training in specifications and codes, structural/mechanical design, client relations and accounting. One can only speculate why this would be the case. Perhaps it is because the midsize firms still hope to use consultants on some aspects of projects, while the small firms try to do it themselves and the large firms have these knowledge area capabilities on staff.

Table 5.5 Areas in which an architect needs to receive training: differences by size of firm * (per cent (no.))

Training area:	Firm size		
	<11 (Small)	11–30 (Midsize)	31+ (Large)
Specifications and codes	92.9 (473)	77.1 (37)	92.3 (12)
Structural/mechanical design	86.5 (443)	57.4 (27)	84.6 (11)
Client relations	81.0 (414)	66.7 (32)	76.9 (10)
Accounting	58.7 (294)	39.1 (18)	58.3 (7)
Real estate development	43.9 (219)	66.7 (32)	66.7 (8)

*   The figures represent the principals, by firm size, agreeing an architect needs to receive training in the area listed.

How well was I trained?

The principals also report differences in the adequacy of their university-level education, as indicated in Table 5.6. It was only in the few subject areas listed in Table 5.6 that there were statistically significant differences among firm sizes (the report of all the areas is elsewhere in this study). Yet, it appears from these data that principals from large firms felt best prepared by their university-level education, with the important exception of preparation in structural and mechanical design.

Table 5.6 Areas in which principals received adequate training: differences by size of firm* (per cent (no.))

Training area:	Firm size		
	11 (Small)	11–30 (Midsize)	31+ (Large)
Building technology	81.5 (424)	64.6 (31)	100.0 (14)
Structural/mechanical design	76.4 (399)	58.3 (28)	73.3 (11)
Urban design and planning	57.8 (301)	63.0 (29)	80.0 (12)
Production	49.0 (236)	50.0 (24)	78.6 (11)
Brief preparation	48.3 (243)	43.5 (20)	71.4 (10)

*   The figures represent the principals, by firm size, agreeing they received adequate training in the area listed.

Attitudes towards the profession and its future

The questionnaire listed a broad variety of statements regarding the principals' view of the future of architecture as a profession. The responding principals were asked to indicate their level of agreement with each statement. As it turns out, very few of these statements show statistically significant differences based on the size of the firm with which the principal directs: see Table 5.7.

Table 5.7 View of the future: differences by size of firm: * (per cent (no.))

	Firm size		
	<11 (Small)	11-30 (Midsize)	31+ (Large)
Statement:			
Computers vital in the future	72.8 (394)	90.0 (45)	78.6 (11)
Workable schedule is a top priority	65.1 (350)	85.4 (41)	100.0 (14)
Competition leads to new approaches	52.0 (281)	72.0 (36)	71.4 (10)
Increased demand gave greater choice	35.9 (195)	56.3 (27)	57.1 (8)
Work outside architecture for more money	24.1 (128)	21.3 (10)	0.0 (0)
Design-only firms are obsolete	21.6 (115)	22.0 (11)	6.7 (1)

*   The figures represent the principals, by firm size, agreeing with a particular statement: see Appendix III for exact wording.

While there is general agreement that computers will be vital to architecture in the future, it appears that principals from midsize firms are more likely to hold this view. Perhaps this could be explained if one assumes that principals from midsize firms have their eyes trained on becoming large firms while principals from small firms have no such ambition.

As firms get larger, it is more likely that principals report that setting a workable time schedule for projects is a top priority for their firm. The same can be said for the statements that greater competition in the architectural field has forced principals to explore new approaches to architectural practice and that the increased demand for architectural services (in the 1980s) gave them greater choice of projects. Thus, it would appear that principals in the larger firms feel more prepared to enter and compete for geographically expanding markets. Also, the larger firms, perhaps through bureaucratization and specialization, have made developing workable time schedules a top priority. While the question was not directly asked, this may imply that the traditional all-night 'charette' most architects learn in university may remain in practice more an artefact of the smaller firms, with larger firms developing a more consistent movement of work.

Principals from smaller firms are more likely to be willing to work outside architecture for more money. Those from the large firms suggest that they would not leave architecture for any amount of money. Also, while there is generally weak belief that firms which focus entirely on design have become obsolete, only one principal from a large firm believed this to be true. The belief appears to be that there is a role for the design-only firm. Such a firm would, of course, work on a project in a subsidiary role. Perhaps principals in large firms recognize the need for such subcontractors while the small and midsize firms, attempting to do it all, are less likely to acknowledge such a role. Unfortunately, the questionnaire did not ask about whether each principal's firm actually does work in a design-only subsidiary role. This could have been very useful information.

Demographics

There are a few demographic items whose differences based on firm size were not statistically significant but are still worth reporting. These are shown in table 5.8. The demographic variables hold little surprise. Larger firms are more likely to employ professionals from other fields and view a wider geographic area for acquiring work. One point of difference is interesting, however.

Unfortunately we cannot say why midsize firms are less likely to employ interior designers and quantity surveyors.

Table 5.8 Demographic items: differences by size of firm, modal response *

	Firm size		
	<11 (Small)	11–30 (Midsize)	31+ (Large)
Statement:			
No. interior designers in firm	<10	0	11–30
No. quantity surveyors in firm	<10	0	11–30
No. engineers in firm	<10	11–30	31+
No. technicians in firm	<10	11–30	31+
Geographic region handled by firm	Regional	National	International

* The figures represent the mode of that firm size. Thus the modal small firm has less than 10 interior designers and the modal large firm has more than 30 technicians.

Conclusions about firm size

Overall, one must conclude, as perhaps one might expect, that the larger firms are better prepared to enter the European and global markets which the profession now faces. They are better organized and have more knowledge areas readily at their disposal to apply to work. Yet, the vast majority of firms are small firms. The cottage industry of the 'gentleman architect' is still quite pervasive in the way the profession operates. Is this at odds with the economic times? Does it augur ill for the future of the profession? For example, many of the tasks performed by architects can be performed by others outside this profession. The profession of architecture is built on a mass of small firms without highly organized marketing strategies (or, more generously, with marketing strategies based more on personal contact than professional expertise). Does this mean that others from outside architecture with greater marketing skills and greater resources at their disposal can attract the market now directed towards the offices of the small architectural firm? While the question can only be definitively answered by the passage of time, the threat is significant and clear.

Differences by building type specialization of the firm

Do principals of firms which specialize in different building types hold different views of the profession of architecture and its future? To answer this, the responses to all questions were compared alongside what was reported as the firm's specialization. A firm's specialization was computed in the following manner. The principals were asked to tick off if their firm spends up to 10 per cent, from 11 per cent to 25 per cent, from 26 per cent to 50 per cent, or more than 50 per cent of its time on housing projects, commercial and industrial buildings, institutional and public buildings, urban design, individual clients or other building types. For purposes of analysis, a firm is considered to have a specialization in one of those areas if the principal reported that the firm spent more than 25 per cent of its time in that area. However, if a principal reported that it spent more than 25 per cent of its time on more than one building type, the firm was coded as having a specialization in that building type for which the highest percentage was reported. If a firm had no building type in which it focused more than 25 per cent of its work, it was coded as mixed and thereby removed from this analysis of the effect of the specialization of the firm.

In fact, it turns out that the specialization of the firm is related to many other issues, as the following discussion shows. Only those differences where $p < 0.05$ are reported here.

Specialization of the firm

It is quickly seen from the data in Table 5.9 that architecture firms do in fact develop specialities. While just under 20 per cent of the principals report practices that are mixed, over a quarter of the firms appear to focus their work on serving individual clients. Only very slightly fewer firms focus on commercial and industrial buildings. About 15 per cent of firms focus on institutional and public clients and about 13 per cent focus on housing projects. Only 1.3 per cent of firms reported that a large portion of their time is spent on urban design. However, this may be an artefact of the sample focusing on private sector firms.

Table 5.9 Specialization of the firm * (per cent (no.))

Specialization area		Excluding 'mixed'
Mixed	19.7 (120)	
Housing	12.6 (77)	15.7 (77)
Commercial and industrial	25.1 (153)	31.2 (153)
Institutional and public clients	14.8 (90)	18.4 (90)
Urban design	1.3 (8)	1.6 (8)
Individual clients	26.6 (162)	33.1 (162)
Total	**100.0 (610)**	**100.0 (490)**

*   Specialization is defined as a firm reporting that it spends over 25 per cent of its time in the named area.

For the remaining analysis in this section, firms which showed no specialization or mixed practices are excluded from the statistical analysis. The intent is to compare whether each listed area of specialization appears to influence other aspects of the profession.

Additionally, so few of those specializing in urban design were detected in the sample that it is difficult to interpret those data. Therefore, the responses of the principals from firms which specialize in urban design are, generally, excluded from the discussion.

The work of the principal

Table 5.10 provides our first definite indication that firms which specialize differ in how they practise. As shown in the table, principals of firms which specialize in housing or urban design are likely to spend more time on building design. Principals of firms which specialize in institutional and public buildings are likely to spend more time on publicizing the firm's work.

Principals of firms who specialize in serving individual clients are most likely to spend time on production drawings.

Principals of firms which specialize in urban design are most likely to spend time co-ordinating consultants and meeting with clients. While it was not a common result, it would appear that principals of firms who serve institutional and public clients or individual clients are significantly more likely to spend time writing specifications.

Table 5.10 Time spent by the principal: differences by building type specialization of firm * (per cent (no.))

| Activity: | Specialization | | | | |
	Housing	Commercial/ industrial	Institute/ public	Urban design	Individual clients
Building design	51.9 (40)	38.6 (59)	15.6 (14)	62.5 (5)	37.0 (60)
Publicizing work	46.8 (36)	58.8 (90)	70.0 (63)	62.5 (5)	53.1 (86)
Production drawings	24.7 (19)	22.9 (35)	22.2 (20)	37.5 (3)	39.5 (64)
Meeting with clients	20.8 (16)	13.7 (21)	3.3 (3)	37.5 (3)	15.4 (25)
Co-ordinating consultants	14.3 (11)	17.6 (27)	21.1 (19)	62.5 (5)	6.8 (11)
Site supervision	11.7 (9)	11.1 (17)	21.1 (19)	50.0 (4)	16.0 (26)
Meeting with project managers	2.6 (2)	0.7 (1)	1.1 (1)	25.0 (2)	1.9 (3)
Writing agreements	1.3 (1)	1.3 (2)	3.3 (3)	25.0 (2)	1.2 (2)
Writing specifications	0.0 (0)	1.3 (2)	4.4 (4)	0.0 (0)	8.0 (13)
Writing construction budgets	2.6 (2)	1.3 (2)	1.1 (1)	25.0 (2)	6.8 (11)

* The figures represent the percentage of principals, by the specialization of principals' firms, reporting that they spend 20 per cent or more of their work week on the listed activity: see Appendix III for exact wording.

Organizing one's work

As shown in Table 5.11, principals of firms that specialize in individual clients are much more likely to agree that the architect should meet separately with the various project participants. Yet, those who focus on housing or commercial and industrial buildings, while still a minority, are more likely to agree that one should seek client approval only after the architect has decided the design issues.

The avoidance of conflict is rated as most important by principals of firms who serve individual clients and specialize in housing projects. While law and liability are reported as important by the majority of responding principals in all categories, principals of firms which specialize in commercial and industrial buildings or institutional and public buildings are more likely to report that this area is important. Principals from all firms with specializations report that it is important to hire new staff with a wide variety of capabilities but there is a statistically significant, if small, difference among the categories. The same is true regarding the importance of hiring new staff with strong design talents; however, for every category, the importance of this was less than for hiring new staff with wide capabilities. The same is true (except it is even less important overall) for the importance of hiring staff with philosophies similar to those already in the firm.

63

Table 5.11 Organizing one's work: differences by building type specialization of firm *
(per cent (no.))

	Specialization				
	Housing	Commercial/ industrial	Institute/public	Urban design	Individual clients
Agreeing with statement:					
The architect should meet separately with each project participant	30.3 (23)	25.3 (38)	31.8 (28)	87.5 (7)	49.7 (79)
Use meetings for client approval only after design issues are decided	32.5 (25)	34.0 (51)	21.1 (19)	14.3 (1)	25.6 (41)
Reporting as important:					
Avoidance of conflict	97.4 (75)	83.6 (127)	79.5 (70)	75.0 (6)	90.6 (145)
Law and liability	93.2 (68)	96.1 (147)	95.5 (84)	75.0 (6)	87.6 (141)
Hiring staff with wide capabilities	86.8 (66)	90.1 (137)	89.2 (74)	87.5 (7)	78.8 (123)
Hiring talented new staff	80.0 (60)	82.8 (125)	77.6 (66)	87.5 (7)	65.8 (102)
Hiring new staff with similar philosophies	73.3 (55)	64.5 (98)	64.6 (53)	37.5 (3)	59.9 (94)
Engineering	70.7 (53)	73.0 (111)	74.7 (65)	50.0 (4)	75.3 (119)
Public recognition	67.5 (52)	58.3 (88)	57.3 (51)	50.0 (4)	43.8 (71)
Technological innovation	65.8 (50)	65.4 (100)	56.3 (49)	75.0 (6)	56.6 (90)
Obtaining client leads from other professionals	45.5 (35)	69.3 (106)	47.7 (41)	25.0 (2)	48.8 (79)
Keeping up with real estate news	39.0 (30)	63.4 (97)	50.0 (43)	50.0 (4)	45.1 (73)
Inviting clients out socially	36.4 (28)	56.6 (86)	54.1 (46)	25.0 (2)	33.5 (54)
Joining client organizations	14.3 (11)	24.2 (37)	24.7 (22)	12.5 (1)	16.4 (26)
Globalization of the Profession	10.6 (7)	25.7 (37)	19.5 (15)	62.5 (5)	13.4 (20)
Working in a subsidiary design role	7.9 (6)	26.1 (40)	18.8 (16)	0.0 (0)	11.5 (18)

* The figures represent the principals, by the specialization of principals' firms, agreeing with or reporting as important a particular approach to work organization: see Appendix III for exact wording.

Interestingly, those in firms which specialize in commercial or industrial projects are most likely to report that architects should be willing to work in a subsidiary, design-only role, perhaps reflecting the nature of these projects.

What training do architects need?

Except for firms which specialize in urban design, there is strong agreement that an architect needs training in building technology, structural/mechanical design and schematic design, even given sometimes small but statistically significant differences, as shown in Table 5.12. Yet, the need for training in computer-aided design and history of architecture is most strongly reported by principals of firms who specialize in commercial or industrial projects or institutional or public projects. Similarly, principals of firms which specialize in serving individual clients, commercial or industrial projects or institutional or public projects are most likely to report that budget management, office management, project management and construction management should be part of an architect's training. Unfortunately, the questionnaire and data do not provide any obvious explanation.

Table 5.12 Areas in which an architect needs to receive training: differences by building type specialization of firm* (per cent (no.))

	Specialization				
	Housing	Commercial/ industrial	Institute/public	Urban design	Individual clients
Training area:					
Building technology	95.8 (69)	99.3 (144)	97.5 (79)	75.0 (6)	96.8 (150)
Structural/mechanical design	78.1 (57)	80.4 (115)	88.9 (72)	100.0 (8)	81.9 (127)
Computer-aided design	82.2 (60)	92.4 (134)	91.3 (73)	75.0 (6)	78.1 (121)
Client relations	74.0 (54)	86.2 (125)	70.9 (56)	75.0 (6)	76.1 (118)
Schematic design	88.9 (64)	100.0 (145)	100.0 (79)	100.0 (8)	94.6 (140)
History of architecture	78.1 (57)	88.3 (128)	92.6 (75)	100.0 (8)	82.1 (124)
Budget management	68.1 (49)	83.4 (121)	84.2 (64)	75.0 (6)	83.4 (126)
Office management	73.0 (54)	85.2 (121)	71.4 (55)	75.0 (6)	87.6 (134)
Project management	69.9 (51)	83.2 (119)	78.9 (60)	75.0 (6)	80.7 (121)
Construction management	75.3 (55)	77.8 (112)	78.8 (63)	62.5 (5)	80.9 (127)

* The figures represent the principals, by the specialization of principals' firms, agreeing an architect needs to receive training in the area listed.

How well was I trained?

It is slightly difficult to interpret the responses on the adequacy of training shown in Table 5.13. First, some of these principals were trained some time ago. So, this certainly should not be viewed as a reflection of today's university-level curricula. Secondly, if a subject received very little coverage in training but was very little needed in practice, then its coverage would still be adequate. These data must be viewed in this light.

Table 5.13 Areas in which principals received adequate training: differences by building type specialization of firm * (per cent (no.))

| Training area: | Specialization | | | | |
	Housing	Commercial/ industrial	Institute/ public	Urban design	Individual clients
Brief preparation	52.7 (39)	44.8 (64)	48.8 (41)	100.0 (8)	52.1 (75)
Facility management	2.8 (2)	7.7 (10)	10.3 (8)	25.0 (2)	10.4 (14)
Specifications and codes	37.8 (28)	57.3 (82)	58.8 (50)	62.5 (5)	48.7 (75)
Construction management	35.1 (26)	24.3 (34)	37.8 (31)	50.0 (4)	29.4 (42)
Research	28.4 (21)	35.0 (49)	41.8 (33)	75.0 (6)	47.6 (68)
Real estate development	10.8 (8)	5.8 (8)	11.3 (9)	50.0 (4)	11.2 (16)
Project management	18.9 (14)	18.6 (26)	23.2 (19)	75.0 (6)	22.5 (32)
Computer-aided design	11.1 (8)	6.5 (9)	13.9 (11)	0.0 (0)	18.4 (25)
Marketing	14.1 (10)	7.1 (10)	5.0 (4)	25.0 (2)	7.7 (11)
Budget management	15.3 (11)	10.1 (14)	16.0 (13)	50.0 (4)	14.3 (20)

* The figures represent the principals, by the specialization of principals' firms, agreeing they received adequate training in the area listed.

Principals from firms which specialize in commercial or industrial buildings or institutional or public projects are more likely to report they received adequate training in specification and codes. But, that seems to limit their positive reporting. Those in firms which specialize in housing or individual clients are most likely to report that they received adequate training in brief preparation. This may reflect a circular causation (chicken and egg syndrome). For example, you may focus on individual clients because you feel well trained in brief preparation; conversely, focusing on individual clients may lead you to believe that your training in brief preparation was adequate. From the questionnaire data we cannot identify which causal direction is correct.

Those serving institutional or public clients and those serving individual clients are most likely to report adequate training in facility management. Those specializing in housing are most likely to report adequate training in construction management. Those specializing in housing or serving individual clients are most likely to report adequate training in real estate development and budget management. Those serving individual clients are most likely to report adequate training in research.

Attitudes towards the profession and its future

Table 5.14 shows that those specializing in commercial or industrial clients are least likely to agree that no matter what happens on a project the final responsibility falls to the architects, and that only a principal should negotiate with a client, perhaps reflecting the complex nature of such projects.

Table 5.14 View of the future: differences by building type specialization of firm * (per cent (no.))

Statement:	Specialization				
	Housing	Commercial/industrial	Institute/public	Urban design	Individual clients
Final responsibility falls to the architect	87.7 (64)	79.9 (119)	87.2 (75)	87.5 (7)	85.0 (136)
Workable schedule is a top priority	55.8 (43)	76.0 (114)	60.5 (52)	87.5 (7)	62.3 (101)
Only the owner and principal determine fees	64.9 (50)	68.7 (103)	81.6 (71)	87.5 (7)	77.2 (125)
Firms must offer comprehensive services	70.7 (53)	85.6 (131)	80.0 (72)	62.5 (5)	74.1 (120)
Private contacts lead to clients	81.3 (61)	87.3 (131)	88.4 (76)	75.0 (6)	93.2 (150)
Impossible to determine tasks before job begins	92.2 (71)	74.3 (113)	78.2 (68)	50.0 (4)	93.2 (150)
Computers vital in the future	77.9 (60)	84.3 (129)	78.2 (68)	75.0 (6)	63.6 (103)
Hard to find 30 minutes of uninterrupted time	77.6 (59)	69.7 (106)	55.8 (48)	62.5 (5)	67.3 (109)
Increased competition leads to new approaches	50.0 (38)	67.3 (103)	56.2 (50)	75.0 (6)	42.2 (68)
Architects should offer design-only services and allow others to lead the team	49.4 (38)	55.0 (83)	33.3 (29)	25.0 (2)	54.0 (83)
Only principal or owner of firm should evaluate work of associates	49.4 (38)	48.7 (72)	66.3 (57)	37.5 (3)	58.4 (94)
Radical ideas are best avoided	50.6 (39)	29.1 (44)	23.0 (20)	0.0 (0)	32.1 (52)
Only principals should negotiate with owner	48.1 (37)	44.1 (67)	46.0 (40)	37.5 (3)	54.9 (89)
Increased demand gave greater choices	37.7 (29)	45.3 (68)	39.3 (35)	37.5 (3)	33.3 (54)
Firms which focus on design are obsolete	23.4 (18)	26.5 (40)	19.8 (17)	00.0 (0)	16.3 (26)

*   The figures represent the principals, by the specialization of principals' firms, agreeing with a particular statement: see Appendix III for exact wording.

They are most likely to agree that firms should offer comprehensive services, that a top priority is developing a workable time schedule, that computers will be vital to the profession in the future, that architects should be willing to offer design- only services and permit others to lead the building team, that increased competition leads them to develop new approaches, that firms which focus only on design are obsolete, and that increased demand for architectural services in the 1980s gave them greater choice of projects.

Those who are principals of firms which specialize in institutional or public buildings are most likely to agree that only the owner or principal should determine the architects' fees, and that only the owner or principal of a firm should evaluate an associate's work. Those who are principals of firms which specialize in serving individual clients are most likely to agree that private contacts lead to clients, and that it is impossible to know all the tasks involved in a project before the project begins (closely followed by those whose firms specialize in housing). Those who are principals of firms which specialize in housing are most likely to agree that it is hard for them to find thirty minutes of uninterrupted time, and that radical ideas are best avoided.

Demographics

In all specialization areas, as shown in Table 5.15, the largest percentage of firms work on projects in the £1 million to £8 million range. Firms with specialities in commercial or industrial buildings are most likely to have projects at about the £8 million line.

Table 5.15 Demographic items: differences by building type specialization of firm * (per cent (no.))

	Specialization				
	Housing	Commercial/ industrial	Institute/public	Urban design	Individual clients
Statement:					
Geographic region of firm's work (median)	Regional	Regional	Regional	National	Regional
Volume of business per year:					
< £0.5 million	15.6 (12)	10.5 (16)	15.6 (14)	12.5 (1)	35.2 (57)
£0.5 – 1 million	6.5 (5)	8.5 (13)	10.0 (9)	12.5 (1)	16.0 (26)
£1 – 8 million	50.6 (39)	41.8 (64)	47.8 (43)	25.0 (2)	42.0 (68)
> £8 million	27.3 (21)	39.2 (60)	26.7 (24)	50.0 (4)	6.8 (11)
In a multi-disciplinary practice	28.6 (22)	52.9 (81)	38.9 (35)	87.5 (7)	21.0 (34)

*   The figures represent the principals, by the specialization of that principals' firms, agreeing with a particular statement.

The firms with urban design specialities are most likely, followed by the firms with a speciality in commercial or industrial projects, to be multi-disciplinary practices.

Conclusions about effects of building type specialization of the firm

While it is clear that the development of specializations within a firm does appear to have a relation with the views of the principal of that firm, these differences are explained by the nature of these projects. Working for individual clients is a very personal activity often involving few consultants or external expertise. Commercial and industrial projects can be extremely complicated involving the co-ordination of many disciplines. Architects who design institutional or public buildings are often responsible for working with different client groups – and large bureaucracies. These differences in the types of project can easily explain most, if not all, of the differences in the views of the principals from firms with different specializations.

Differences by multi-disciplinary practice

It is a commonly held truth (or perhaps myth) that firms are getting larger and that they are becoming increasingly multi-disciplinary. Comparable data are not available to identify whether this increase is true. But, our sample showed that 35.4 per cent (216) of the principals are in firms which should be described as multi-disciplinary. A firm was categorized as multi-disciplinary if it employed architects and at least one member of the allied professions: interior designers, planners, quantity surveyors, engineers, and landscape architects. Thus, over a third of architecture firms in the United Kingdom are currently multi-disciplinary. Can it be said that these firms operate differently from the single-discipline firms? Do principals from multi-disciplinary firms have views of the profession of architecture that differ from principals in architect-only firms? It is these questions that this section addresses. All differences that are reported were statistically significant differences with $p < 0.05$.

Specialization of the firm

As shown in Table 5.16, multi-disciplinary practices are more likely to do work in the commercial and industrial sector and less likely to do work in the housing sector or work for individual clients (which is also most likely to be housing). This, of course, makes good sense.

Table 5.16 Specialization of the firm: *differences of multi-disciplinary practices (per cent (no.))

	Multi-disciplinary?	
	No	Yes
Specialization area:		
Housing	50.3 (148)	31.3 (56)
Commercial and industrial	43.3 (133)	55.5 (111)
Individual clients	46.9 (160)	21.9 (39)

*   Specialization is defined as a firm reporting that it spends over 25 per cent of time on the named area.

Commercial and industrial projects are more complex than housing projects (or at least are perceived as such) and therefore require greater expertise than the architect alone can bring to the project. The multi-disciplinarity that the firm can bring to such a project would be likely to increase the probability of a firm being awarded the job.

The work of the principal

Does the work of principals in multi-disciplinary firms vary from the work of principals in architect-only firms? It would appear from the responses of the principals in Table 5.17 that, yes, in fact it does.

Table 5.17 Time spent by the principal: differences of multi-disciplinary practices * (per cent (no.))

	Multi-disciplinary?	
Activity:	No	Yes
Building design	36.3 (143)	32.9 (71)
Production drawings	31.0 (122)	20.8 (45)
Site supervision	15.0 (59)	32.9 (71)
Co-ordinating consultants	10.2 (40)	20.4 (44)
Writing specifications	3.8 (15)	2.8 (6)
Recruiting clients in person	2.3 (9)	3.7 (8)
Meeting with project managers	1.8 (7)	2.8 (6)
Staffing office	1.5 (6)	0.5 (1)
Developing office procedures	0.3 (1)	1.9 (4)

*   The figures represent the percentage of principals, by whether their practice is multi-disciplinary, reporting that they spend 20 per cent or more of their work week on the listed activity: see Appendix III for exact wording.

Principals of multi-disciplinary firms are less likely to spend time on building design, production drawings, writing specifications, and office staffing. They are more likely to spend time on site supervision, co-ordinating consultants, recruiting clients in person, meeting with project managers and developing office procedures. In effect, principals of multi-disciplinary firms are more likely to spend time on management-related activities, both the management of the firm as well as the management of the project.

Organizing one's work

Table 5.18 demonstrates that principals of multi-disciplinary firms are more likely to agree that preliminary meetings in a project are the most crucial, and that the following factors are important: land usage in design, hiring talented new staff, technological innovation, obtaining leads on clients from other professionals, inviting clients out socially, telephoning clients informally, the increasing cultural diversity of clients, joining the organizations that clients join, the globalization of the profession and the willingness to work in a design-only subsidiary role.

Table 5.18 Organizing one's work: differences of multi-disciplinary practices * (per cent (no.))

	Multi-disciplinary?	
	No	Yes
Agreeing with statement:		
Preliminary meetings are most crucial	90.4 (349)	93.4 (197)
Reporting as important:		
Avoidance of conflict	90.2 (349)	82.6 (176)
Professional code of ethics	90.1 (355)	83.2 (178)
Land usage in design	71.5 (269)	80.7 (171)
Hiring talented new staff	68.4 (255)	87.9 (188)
Technological innovation	59.2 (226)	72.2 (153)
Getting leads on clients from other professionals	47.0 (181)	63.4 (135)
Inviting clients out socially	35.4 (134)	56.1 (119)
Calling clients Informally	35.1 (135)	56.9 (120)
Cultural diversity of clients	31.2 (119)	41.0 (87)
Joining client organizations	13.5 (52)	27.2 (58)
Globalization of the profession	13.4 (47)	25.0 (49)
Working in a subsidiary design role	12.7 (48)	21.2 (45)

* The figures represent the principals, by whether their practice is multi-disciplinary, agreeing with or reporting as important a particular approach to work organization: see Appendix II for exact wording.

If there is a pattern, it is that principals of multi-disciplinary firms are more conscious of the marketing of their services. They are more likely to see the activities that would be considered marketing efforts as important. As discussed elsewhere, it is the larger firms, which are also more likely to be the multi-disciplinary firms, which appear to have organized marketing efforts. These principals from multi-disciplinary firms are less likely to report that the avoidance of conflict and the professional code of ethics are important. Again, perhaps because of greater concentration on management skills and marketing activities, these two items become less important because the factors they represent are handled in other ways.

What training do architects need?

There were only four areas of training in which the principals of multi-disciplinary firms differed from the principals of architect-only firms, as indicated in Table 5.19. In all four cases, the principals of multi-disciplinary firms were more likely to report that an architect needs to receive training in these areas. These are schematic design, computer-aided design, production drawings and research.

Table 5.19 Areas in which an architect needs to receive training: differences of multi-disciplinary practices * (per cent (no.))

	Multi-disciplinary?	
	No	Yes
Training area:		
Schematic design	93.4 (339)	98.0 (196)
Computer-aided design	83.9 (313)	88.5 (177)
Production drawings	74.6 (265)	83.7 (164)
Research	58.6 (212)	67.7 (134)

* The figures represent the principals, by whether their practice is multi-disciplinary, agreeing an architect needs to receive training in the area listed.

Again, perhaps because the multi-disciplinary practices are more likely to be larger firms, permitting individual specialization of tasks, and perhaps because these larger firms are more likely to be computerizing and looking towards the longer-range future, these items were more likely to be reported as needed in the training of an architect. One certainly cannot accuse the multi-disciplinary firms of ignoring the role of design in the architect's work.

How well was I trained?

Table 5.20 shows that the principals of multi-disciplinary firms report only three areas where they differ from principals of architect-only firms in how adequately they believe they were trained. They are more likely to report that they received adequate training in schematic design, brief preparation and production drawings.

Table 5.20 Areas in which principals received adequate training: differences of multi-disciplinary practices* (per cent (no.))

	Multi-disciplinary?	
	No	Yes
Training area:		
Schematic design	82.8 (309)	89.9 (187)
Brief preparation	46.1 (165)	52.7 (108)
Production drawings	44.6 (153)	58.7 (118)

*   The figures represent the principals, by whether their practice is multi-disciplinary, agreeing they received adequate training in the area listed.

Attitudes towards the profession and its future

Regarding their view of the profession in general and the future of the profession, the principals of multi-disciplinary practices revealed a number of topics on which they differed from the principals of architect-only firms, as shown in Table 5.21. They are more likely to agree that architect and clients should reach agreement on projects early in the schematic design stages; that computers will be vital to the future practice of architecture; that developing a workable time schedule is a top priority; that increased competition has led to new approaches; that increased demand for architectural services in the 1980s gave them a greater choice of projects; and that efficiency is more important than creativity in design.

They also are less likely to agree that private contacts lead to clients; that it is impossible to predict what tasks a job entails before a project begins; that architects adhere to a strict code of ethics; that architects should offer design-only services and allow others to lead the building team; that negotiations should be done only by a firm's principals; that radical ideas are best avoided; and that they would work outside architecture for more money.

The pattern here adds confirmation to the pattern seen above. The multi-disciplinary firms (which also tend to be larger firms) would appear to be more focused in management and marketing approaches; more directed towards schedules, efficiency and productivity; more comfortable in facing competition; and more likely to offer a wider range of services.

Table 5.21 View of the future: differences of multi-disciplinary practices * (per cent (no.))

	Multi-disciplinary?	
	No	Yes
Statement:		
Architect and client should reach agreement early in schematic design	92.3 (362)	96.8 (209)
Private contacts lead to clients	89.7 (347)	81.5 (172)
It is impossible to determine tasks before a project begins	89.5 (350)	71.2 (151)
Computers vital in the future	70.7 (278)	81.1 (172)
Architects adhere to a strict code of ethics	63.0 (243)	49.3 (104)
Workable time schedule is a top priority	62.0 (241)	77.7 (164)
Architects should offer design-only services and allow others to lead the team	52.4 (205)	42.2 (89)
Negotiation with owners should be done by principals only	53.1 (207)	41.3 (88)
Increased competition led to new approaches	45.9 (178)	69.0 (149)
Radical ideas are best avoided	36.8 (143)	22.6 (48)
Increased demand (in 1980s) gave me greater choice of projects	34.3 (134)	44.9 (96)
Work outside architecture for more money	27.6 (108)	24.2 (52)
Efficiency is more important than creativity in design	27.0 (104)	28.2 (60)

*   The figures represent the principals, by whether their practice is multi-disciplinary, agreeing with a particular statement: see Appendix III for exact wording.

Demographics

The principals in multi-disciplinary firms are more likely to serve a national clientele; likely to have been in business as a firm for longer; and more likely to undertake larger projects.

Conclusions about effects of multi-disciplinary practices

Overall, one must conclude that the principals of the multi-disciplinary firms report themselves as more prepared and more comfortable in facing the future of their profession.

Table 5.22 Demographic items: differences of multi-disciplinary practices (per cent (no.))

Statement	Multi-disciplinary?	
	No	Yes
Geographic region served (median)	Regional	National
Years firm in existence (median)	11–20 years	20–100 years
Volume of business per year :		
< £0.5 million	25.9 (102)	18.5 (40)
£0.5 – 1 million	16.2 (64)	3.7 (8)
£1– 8 million	44.4 (175)	38.0 (82)
> £8 million	13.5 (53)	39.8 (86)

Differences by age group

Specialization of the firm

A firm was considered to have developed a specialization in a particular area if it spends more than 25 per cent of its time conducting work in that area of expertise. Extremely few principals reported a specialization in urban design. Yet, those above age 40 were nearly twice as likely to specialize in urban design as those below 40 years old. The 40 to 49 age group was more likely than those younger or those older to have developed specializations in housing or commercial and industrial buildings and less likely to work on public or institutional buildings or for individual clients. The data in Table 5.23 appear to be asking a very interesting question: do firms change their areas of specialization when the principals are in the 40 to 49 age range, only to return to the earlier areas of specialization after the principal reaches 50 years old?

Table 5.23 Specialization of the firm :* differences by age of principal (per cent (no.))

Specialization area:	Age of principal		
	< 40	40–49	50–65
Housing	39.5 (30)	46.0 (97)	79.8 (66)
Commercial and industrial	48.4 (45)	50.0 (108)	46.9 (84)
Institutional and public	30.4 (17)	29.9 (46)	38.7 (46)
Urban design	5.6 (1)	4.8 (4)	9.1 (4)
Individual clients **	54.4 (49)	30.1 (63)	35.3 (67)

*   Specialization is defined as a firm reporting that it spends over 25 per cent of its time on the named area.
**   Only this difference is statistically significant at p <0.05.

Do these data indicate that firms start out with individual clients and public and institutional buildings; then, as the firm grows, more housing and commercial or industrial projects are taken on, reducing the focus on individual and public or institutional clients; then, in the final stage, there is some return to institutional and public buildings and a reduction in housing, commercial and industrial building? It is impossible to answer this question without longitudinal data on a panel of responding principals. Perhaps this progression only reflects the emphases common in the educational curriculum at the time an individual principal attended university. Yet, these are the first data known to these authors indicating that such shifts in the focus of firms may be occurring.

The work of the principal

The principals spend their time at work on a variety of tasks. With the assumption that if a principal spends 20 per cent or more of his or her time (one day per week in total) on a particular task, then that task is an important one, the data indicate that there are statistically significant differences among the age groups: see Table 5.24.

Table 5.24 Time spent by the principal: differences by age of principal * (per cent (no.))

	Age of principal		
	<40	40–49	50–65
Activity:			
Building design	40.0 (42)	34.0 (83)	34.2 (76)
Meeting with clients	10.5 (11)	18.0 (44)	9.0 (20)
Site supervision	9.5 (10)	17.6 (43)	15.3 (34)
Meeting with project managers	3.5 (4)	1.6 (4)	1.8 (4)
Writing agreements	1.9 (2)	4.5 (11)	1.4 (3)
Writing specifications	1.0 (1)	2.9 (7)	4.5 (10)
Negotiating new work	1.9 (2)	1.6 (4)	0.9 (2)
Staffing office	1.9 (2)	0.0 (0)	2.3 (5)
Recruiting clients by mail	0.0 (0)	2.9 (7)	0.9 (2)
Recruiting clients in person	0.0 (0)	3.3 (8)	3.2 (7)
Writing construction budgets	0.0 (0)	2.0 (5)	8.1 (18)
Recruiting clients by telephone	0.0 (0)	1.2 (3)	1.8 (4)
Developing office procedures	0.0 (0)	2.0 (5)	0.0 (0)
Establishing fee structure	0.0 (0)	1.6 (4)	0.0 (0)

* The figures represent the principals, by age group, reporting that they spend 20 per cent or more of their work week on the listed activity: see Appendix III for exact wording.

Principals below 40 years old spend slightly more time working on building design (which includes schematic design and production drawings). Yet it is clear that, whatever the age group, only about 35 to 40 per cent of the principals spend 20 per cent or more of their time on this aspect of practice. The remainder spend less than one day per week on building design.

For the remainder of the items in Table 5.24, only small percentages of the responding principals spend one day per week (20 per cent) or more on these items. Yet, one can see that some age groups are more likely than others to do this for each item listed. For example, those in the middle group (age 40 to 49) are more likely to spend at least a day per week meeting with clients, in site supervision, writing agreements, recruiting clients by mail and in person, and developing office procedures.

Those in the less than 40 age group are more likely to spend one day or more per week meeting with project managers and negotiating new work than those in the two older age groups. Principals in the 50 and above group are most likely to spend one day per week or more writing specifications, staffing the office, writing construction budgets, and recruiting clients by telephone. As with all age-related factors, it is always risky to claim that a pattern has been determined in data like these. The effects of age could be for a variety of reasons. For example, differences due to age could be because experience has led to knowledge and a propensity to concentrate on certain aspects of the task. It, however, could also be that principals of different ages graduated from university at different times. Alternatively, these differences might reflect life-cycle changes: starting out, building up a practice, protecting market share until retirement. None the less, it does appear that those in the oldest group spend more time on the more legalistic aspects of being an architect. Those in the youngest group spend the most time on building design, working with project managers and negotiating new work. Those in the middle group spend the most time publicizing their efforts.

Organizing one's work

Overall, one can see from the data in Table 5.25 that experience generally does appear to lead to wisdom regarding the operation of one's practice. The oldest group of principals are more likely to agree with such statements of general business practice as: preliminary meetings are the most crucial, agreement on schematic design should be reached early, professional codes of ethics are important, informal meetings are more important than formal meetings, calling on clients informally is important, joining client organizations is important, being willing to work in a design-only subsidiary role is important, and project management is important.

Table 5.25 Organizing one's work: differences by age of principal * (per cent (no.))

	Age of principal		
	<40	40–49	50–65
Agreeing with statement:			
Architect and client should reach agreement early in schematic design	84.8 (89)	94.2 (229)	97.3 (215)
Preliminary meetings are most crucial	83.5 (85)	91.7 (222)	96.3 (208)
It is difficult to find 30 minutes of uninterrupted time	69.2 (72)	73.1 (177)	59.6 (130)
The architect should meet separately with each project participant	42.9 (45)	29.7 (71)	41.9 (90)
Informal meetings are more rewarding	32.0 (33)	41.9 (101)	57.1 (120)
Reporting as important:			
Law and liability	87.6 (92)	95.1 (231)	91.2 (197)
Professional code of ethics	81.9 (86)	84.4 (206)	92.8 (206)
Hiring staff with wide capabilities	93.2 (96)	88.8 (215)	76.2 (160)
Land usage in design	72.4 (76)	76.8 (185)	73.9 (156)
Education of client and public	71.4 (75)	84.7 (205)	75.2 (167)
Project management	62.9 (66)	77.9 (190)	83.0 (181)
Calling clients informally	35.2 (37)	39.3 (95)	49.5 (106)
Cultural diversity of clients	27.4 (29)	42.6 (103)	28.2 (60)
Joining client organizations	14.3 (15)	16.3 (39)	23.9 (52)
Globalization of the profession	10.2 (10)	21.8 (50)	15.8 (30)
Working in a subsidiary design role	11.4 (12)	14.9 (36)	19.8 (42)

* The figures represent the principals, by age group, agreeing with or reporting as important a particular approach to work organization: see Appendix III for exact wording.

Yet, interestingly, there are a number of factors which are more likely to be reported as important by the 40 to 49 year old principals. The 40 to 49 age group views law and liability, hiring staff with a wide range of capabilities, land usage in design, the role of the architect in educating the client and the public about architecture, the cultural diversity of clients, and the globalization of the profession as more important than the other two groups. Why? Perhaps members of this middle group are feeling the passage of time in their lives and are seeking to expand their abilities before entering the older age group. Perhaps they are seeing the changing context of practice while the older group, eyeing retirement, are not looking as far ahead.

What training do architects need?

There are only a few statistically significant differences in the views of the topics in which the principals report an architect should receive training. For all the six topics listed in Table 5.26, it is the central age group which is most likely to report an architect needs training in that topic. This may, more than anything else, reflect an ageing process for this group. Do they see this middle period as the time during which they must make a major push in their careers before they reach the age of 50? Is this why they are most likely to report that an architect needs training in areas (except for brief preparation) which are usually not part of an architect's education? Alternatively, are they concerned about the next generation of architects whom they will supervise?

Table 5.26 Areas in which an architect needs to receive training: differences by age of principal * (per cent (no.))

	Age of principal		
	<40	40–49	50–65
Training area:			
Brief preparation	92.0 (92)	96.6 (225)	90.2 (184)
Communication	90.0 (90)	97.4 (227)	87.0 (180)
Office management	79.0 (79)	83.6 (189)	82.8 (168)
Budget management	77.0 (77)	87.2 (198)	78.5 (161)
Accounting	54.1 (53)	62.8 (142)	50.2 (102)

* The figures represent the principals, by age group, agreeing an architect needs to receive training in the area listed.

How well was I trained?

The youngest group of principals is more likely to report that it was adequately trained in schematic design, brief preparation, urban design and planning, communication, and computerization. The oldest group analysed (those between 50 and 65 years old) is most likely to report that they were adequately trained in building technology, interior design, specification and codes, and marketing. The middle group is most likely to report that it was adequately trained in brief preparation, human behaviour, research, and facility management.

Schools of architecture certainly go through phases in their curricula where certain subjects are emphasized and attention to other topics wanes. The middle group, most likely educated in the 1960s or early 1970s received a substantial dose of information about human behaviour in buildings and research. The youngest group's responses may only reflect that they left university during a period when style and form were being emphasized. The oldest group's responses may simply indicate that they were in school during a period when more practical aspects such as specifications, codes and building technology were under the spotlight. Table 5.27 provides the details.

Table 5.27 Areas in which principals received adequate training: differences by age of principal * (per cent (no.))

	Age of principal		
	<40	40–49	50–65
Training area:			
Schematic design	91.0 (91)	88.6 (209)	79.5 (171)
Building technology	76.7 (79)	76.4 (178)	84.5 (180)
Urban design and planning	59.2 (61)	61.6 (143)	53.1 (112)
Interior design	38.8 (40)	49.8 (116)	58.0 (120)
Brief preparation	59.4 (60)	53.0 (123)	40.1 (81)
Specifications and codes	38.8 (40)	44.4 (103)	58.6 (123)
Human behaviour	42.7 (44)	49.1 (114)	29.6 (59)
Communication	47.6 (49)	37.5 (87)	19.1 (39)
Research	41.0 (44)	48.5 (110)	35.2 (70)
Computer-aided design	16.7 (17)	11.0 (24)	9.7 (18)
Computerization	10.0 (10)	9.0 (20)	5.4 (10)
Facility management	6.4 (6)	12.0 (26)	9.6 (18)
Marketing	7.8 (8)	7.5 (17)	10.2 (20)

*   The figures represent the principals, by age group, agreeing they received adequate training in the area listed.

Attitudes towards the profession and its future

As shown by Table 5.28, a high percentage of architects would choose to become architects again if starting over. Yet, the older the principal, the less likely this holds true. Also this older principal is most likely to report that architects should offer design-only services and be willing to let others lead the building team; that efficiency is more important than creativity when designing a project; and that radical ideas are best avoided (perhaps reflecting a conservatism commonly associated with this age group).

The younger the principal, the more likely he or she is to express flexibility in his or her approach to practice. For example, the younger principal is more flexible about who determines the architect's fee. This group is most like to see a design-only firm as a viable alternative and, although the numbers for all three groups are very small, this youngest group is most likely to report that architects always place the public interest above client satisfaction, reflecting, perhaps, the greater idealism often associated with youth.

Perhaps, again, reflecting a striving to achieve a level of accomplishment during their 40s, the middle group is most likely to report that computers will be vital to the future of architecture. This group, possibly because of the stage they have reached in their life-cycle, is the most likely to be willing to work outside of architecture for more money.

Table 5.28 View of the future: differences by age of principal * (per cent (no.))

	Age of principal		
	<40	40–49	50–65
Statement:			
Would become an architect again	77.1 (81)	70.0 (170)	63.8 (141)
Only the owner and principal determine fees	61.5 (64)	75.8 (182)	81.4 (179)
Computers vital in the future	68.0 (70)	84.0 (205)	73.1 (160)
Architects should offer design-only services and allow others to lead the team	46.2 (48)	50.8 (123)	51.4 (112)
Efficiency is more important than creativity in design	20.4 (21)	21.2 (51)	34.6 (75)
Radical ideas are best avoided	23.1 (24)	27.4 (66)	41.1 (90)
Work outside architecture for more money	22.3 (23)	28.2 (68)	20.9 (44)
Design-only firms are obsolete	14.3 (15)	22.5 (54)	22.7 (50)
Architects place public interest above client satisfaction	7.7 (8)	5.0 (12)	6.5 (14)

*   The figures represent the principals, by age group, agreeing with a particular statement: see Appendix III for exact wording.

Demographics

Very few of the demographic questions at the end of the questionnaire showed statistically significant differences when compared across the three age groups. Of course, year of graduation, the number of firms the principal has been employed by, the number of years the principal has been a registered architect, how long the firm has been in existence, and the sex of the principal responding, all, as one would expect, showed statistically significant differences. (The few women found by the sample were all in the two younger age groups.)

Yet, two other demographic variables did show differences according to the age of the principal. All of the firms were most likely to focus on small-scale private buildings or a mixed practice of all types of buildings. But, it was the middle age group which was even more likely to focus on this mixed practice than the other age groups.

Additionally, as one might expect, the youngest group was most likely to work on smaller projects (projects worth less than £0.5 million) while the middle group, perhaps again reflecting an aspiring period of life, were most likely to work on the largest of projects (those worth over £8 million). Table 5.29 provides the details.

Table 5.29 Demographic items: differences by age of principal * (per cent (no.))

	Age of principal		
Statement	**<40**	**40-49**	**50-65**
Primary focus of the firm:			
Corporate	11.9 (12)	10.3 (25)	11.6 (25)
Public sector	3.0 (3)	6.6 (16)	8.8 (19)
Small-scale private	29.7 (30)	17.3 (42)	23.7 (51)
Mixed	55.4 (56)	65.8 (160)	55.8 (120)
Volume of business per year:			
< £0.5 million	21.9 (23)	16.8 (41)	24.3 (54)
£0.5–1 million	16.2 (17)	11.5 (28)	12.2 (73)
£1–8 million	45.7 (48)	41.8 (102)	42.3 (94)
> £8 million	16.2 (17)	29.9 (73)	21.2 (47)

*   The figures represent the principals, by firm size, agreeing with a particular statement.

Conclusions about age

A pattern is emerging here (as well as in the report on differences according to the size of the principals' firms reported elsewhere) that the practice of architecture appears to change according to life cycle and firm size. When the architect is young and in a small firm, he or she appears to be more willing to take risks and to be less concerned about matters like law and liability. Then, as the firm enters the midsize range of firms or the architect enters the middle period of his or her career, there appears to be a push for greater concentration on publicizing one's work and taking on larger projects. It is also this middle group which reports higher educational needs for architects. As the architect ages further, he or she seems to settle into a more conservative pattern with greater focus on risk and the business aspects of the profession. If this pattern is true, it may reflect the general life cycle of the individual as much as any other significant factor. This may also give an indication that continuing professional development courses should be focused towards the middle and older age groups.

Differences of architects specializing in management

Two questions which quickly arise when examining the questionnaire results are:

1. whether one can identify some architects who have already begun to specialize in management tasks; and
2. whether such architects hold different views from principals of similar sized firms who do not specialize in this way.

It turns out that in both cases the answer is yes. In order to do this from the data available from the questionnaire, two critical assumptions needed to be accepted. First, one must accept the notion that only the small firms (e.g. firms with ten or fewer architects) were worth examining. It could be true that a principal from any size firm might answer that he or she was specializing in management activities. (It turns out from the data that 70 principals from firms with ten or fewer architects and 30 principals from firms with eleven or more architects spend 80 per cent or more of their time on management activities.) However, in a large firm there could be many principals and the responding principal could be personally specializing in management activities but only because the firm had a strong internal division of labour and not because new forms of practice require more of these skills to be applied, which is of course our concern. Thus, the smaller firms were separately pulled aside for examination. There were 545 such firms with ten or fewer architects.

The second assumption that was made for the analysis was that if the principal of a small firm reported spending 80 per cent or more of his or her time on any combination of management activities, his role was that of an architect specializing in management. These management activities included: co-ordinating consultants and staff, meeting with clients, meeting with project managers, putting agreements in writing, writing specifications, developing construction budgets, estimating work remaining in projects, other project management activities, publicizing work, working with marketing specialists or public relations consultants, recruiting clients, negotiating new work, staffing the office, developing office procedures and fee structures, managing office finances or other office maintenance duties.

Lastly, to examine whether principals who specialized in management held different views from those of principals of other small firms, a statistical test was run comparing them against all other variables on the questionnaire. Only those comparisons where a statistically significant difference ($p < 0.05$) was found are reported here.

Of the 545 firms with ten or fewer architects, 70 principals (12.8 per cent) appeared to specialize in management (as defined above). The remaining 475 principals from small firms (87.2 per cent) could not be said to spend 80 per cent or more of their time on management activities. Thus it can be said, given the above assumptions, that there is in fact a small sector of architects who have begun to specialize in the management aspects of their task. Also, their views do differ from the views of principals in similarly sized firms in some very specific ways.

Building type specialization of the firm

A firm was considered to have developed a specialization in a particular building type if it spent more than 25 per cent of its time conducting work in that area of expertise. Firms with principals who specialize in management were less likely to work for individual clients. Only 21.7 per cent of the principals who specialize in management reported that 26 per cent or more of the firm's time was spent on work for individual clients. On the other hand, 43.1 per cent of other principals reported that 26 per cent or more of the firm's time was spent on projects for individual clients.

The work of the principal

Principals who specialize in management were more likely to report that it is important to join client organizations; to obtain client leads from consultants; to keep up with real estate news; to personally visit or informally call potential clients; and to invite clients out socially.

Table 5.30 The work of the principal: differences by specialization in management (per cent (no.))

	Principal specializes in management?	
	No	Yes
Rating as important:		
Joining client organizations	14.9 (69)	22.9 (16)
Getting lead on clients from consultants	48.5 (226)	67.6 (46)
Keeping up with real estate news	44.5 (207)	73.5 (50)
Personally visiting potential clients	41.9 (193)	60.9 (42)
Informally calling potential clients	38.7 (179)	59.4 (41)
Inviting clients out socially	34.6 (159)	61.8 (42)
Promoting architectural theory	70.9 (332)	56.5 (39)

*   Ratings of items by principals who specialize in management compared with principals who cannot be said to specialize in management: see Appendix II for exact wording.

All of these are variations on the role of marketing within an architectural firm. It would appear that architects who specialize in management, perhaps not surprisingly, are more likely to focus on marketing activities.

Also, interestingly, these people are less likely to see the promotion of architectural theory as important. Table 5.30 gives the details.

What training do architects need?

Principals specializing in management are less likely to agree that architects should receive training in office management, accounting, and specifications and codes (see Table 5.31). They are more likely to agree that architects should receive training in interior design and computerization.

Table 5.31 Areas in which an architect needs to receive training: differences by whether the principal specializes in management * (per cent (no.))

	Principal specializes in management?	
	No	Yes
Training area:		
Office management	83.6 (366)	74.6 (50)
Accounting	60.1 (261)	49.3 (33)
Interior design	74.0 (328)	89.2 (58)
Specifications and codes	93.9 (416)	86.4 (57)
Computerization	67.0 (293)	86.4 (57)

*   Number of principals who agreed an architect should receive training in the area listed, separated by principals who specialize in management compared with principals who cannot be said to specialize in management.

While this cannot be definitively demonstrated by the data from the question-naire, these responses raise interesting questions for future study. For example, do the principals who specialize in management say an architect does not need training in the obvious management areas (accounting, office management, etc.) because they, as enthusiasts, hire other professionals to handle these matters for them? If this is so, in what are they actually specializing? Indeed, these would be interesting questions.

How well was I trained?

Interestingly, as shown in Table 5.32, those principals who specialize in management feel better trained in structural/mechanical design and more poorly trained in building technology and research than principals of small firms generally. Perhaps this is because they, having specialized in their firms, are more likely to recognize deficits or gaps in their university-level education. This, of course, is a question for future research. Here, it can only remain conjecture.

Table 5.32 Areas in which principals received adequate training: differences by whether they specialize in management* (per cent (no.))

	Principal specializes in management?	
	No	Yes
Training area:		
Building technology	83.3 (378)	69.7 (46)
Structural/mechanical design	75.7 (346)	81.5 (53)
Research	42.3 (181)	35.4 (23)

* Number of principals who agreed an architect should receive training in the area listed, separated by principals who specialize in management compared with principals who cannot be said to specialize in management.

Attitudes towards the profession and its future

There appear to be a few interesting differences between principals who specialize in management and principals from the remaining small firms, as shown in Table 5.33.

Table 5.33 View of the future: differences by whether the principal specializes in management * (per cent (no.))

	Principal specializes in management?	
	No	Yes
Statement:		
Cannot find 30 minutes of uninterrupted time	65.7 (309)	54.4 (37)
Creating workable time schedules are a top priority	64.1 (302)	71.6 (48)
Fees should only be determined by the owner or principal of the firm	76.6 (361)	69.1 (47)
Increased competition leads to new approaches	49.6 (233)	68.6 (48)
Computers will be vital in the future	71.2 (337)	83.8 (57)
I would become an architect again	70.6 (332)	56.5 (39)
I would move out of architecture for more money: – agree	28.5 (134)	20.0 (14)
– agree and don't know combined	41.2 (194)	47.0 (33)

* The figures represent the principals, by whether they specialize in management, agreeing with a particular statement: see Appendix III for exact wording.

Principals who specialize in management are less likely to report that they cannot find 30 minutes of uninterrupted time and are more likely to report that creating workable time schedules are a top priority. They are more likely to report that increased competition leads to new approaches, and that computers will be more vital in the future of architecture. They are less likely to report that only the owner or principal of a firm should determine fees.

All of these tendencies reflect the focus on management activities and are, therefore, not surprising. While they are, however, less likely to move out of architecture for more money, they are less likely, if they had it to do over, to become an architect again, perhaps reflecting a recognition that the management expertise they have developed might be more rewarding in another line of endeavour. Interestingly, if one combines those who are undecided with those who would leave architecture for more money, one finds that there is actually a slightly higher tendency among principals with a management specialization to leave architecture for more money – again, perhaps, reflecting a recognition of the value of their skills in many areas.

Demographics

The demographic variables in Table 5.34 (overleaf) provide additional confirmation that these architects are focusing on management issues. Their firms are more likely to have more architects, interior designers, planners, landscape architects and technicians. Therefore, they are also more likely to be multi-disciplinary in their approach to problems. Their firms are more likely to do work for the corporate client and the public sector, perhaps reflecting a management approach which is more compatible with these client groups. They are also more likely to undertake the most expensive range of projects, even though they are still small firms, again perhaps reflecting a management approach compatible with clients who undertake such large projects. They are also more likely to have been a registered architect longer.

Conclusions about principals who specialize in management

While there are some interesting differences between principals who specialize in management and principals of other firms, these differences actually provide no great surprises. In fact, perhaps what these differences provide is confirmation that these architects are focusing on management issues and knowledge areas. They are consistently more likely to report that management-related subjects are more important to them in their practice of architecture, indicating that specializing in management is a viable, if minority, option for the practice of architecture.

Table 5.34 Demographic items: differences by whether the principal specializes in management * (per cent (no.))

	Principal specializes in management?	
Statement:	No	Yes
No. architects in firm:		
1–5	85.7 (407)	62.9 (44)
6–10	8.0 (38)	31.4 (22)
No. interior designers in firm:		
1–5	15.8 (75)	28.6 (20)
No. planners in firm:		
1–5	8.8 (42)	20.0 (14)
No. landscape architects in firm:		
1–5	5.5 (26)	15.7 (11)
No. Technicians in firm:		
1–5	54.1 (257)	62.9 (44)
6–10	4.4 (21)	7.1 (5)
11–30	2.1 (10)	5.7 (4)
Primary focus of the firm:		
Corporate	7.1 (33)	20.0 (14)
Public sector	5.0 (23)	11.4 (8)
Small-scale private	29.4 (136)	7.1 (5)
Mixed	58.4 (270)	61.4 (43)
Volume of business per year:		
<£0.5 million	20.0 (95)	12.9 (9)
£0.5–1 million	11.2 (53)	0.0 (0)
£1–8 million	53.7 (255)	51.4 (36)
>£8 million	15.2 (72)	35.7 (25)
Years as a registered architect:		
6–10 years	41.3 (196)	58.6 (41)

* The figures represent the principals, by whether they specialize in management, agreeing with a particular statement.

Notes

1. *The Architect and His Office*, London, RIBA, 1962.

Part Three

Types of Architectural Practice

6

Issues for practices

The process of change

The increasing exposure of architectural practice to market forces has led to a shift away from the architect as team leader, the growth of varied specialisms and the increased importance of management techniques to help firms adapt to rapidly changing circumstances. The seven cases that follow offer some specific examples of the ways in which such changes have affected individual firms, how these firms have responded to them and what some of the individuals feel about the changes. Five of the cases describe private sector firms, as this side of the business of architectural design is the focus of the present study. To give a perspective to this emphasis, the two other cases describe, firstly, a local authority department which has attempted to adapt to changing public expectations, and, secondly, a commercial design–build organization with a very different standing in the construction procurement market.

The information gathered in these case studies covers a period of over twenty years, from 1968 to 1990, for which relevant records and accurate memories are not consistently available, particularly for the earlier part of the period. None the less we have made the assumption that principals' memories, and whatever records exist of past periods, can give a sufficient picture to map some useful aspects of the experience of a firm. The practices have all been helpful in contributing information and giving time for interviews for which we are very grateful.

It should be noted that the majority of the research was undertaken during 1991, when the recession of the early 1990s had not taken its full effect, and certainly before any signs of recovery were evident. Subsequent changes have occurred in all the case study organizations. In particular by the time of preparing these studies for this book (mid-1993) most were considerably smaller in staff numbers than at the time of interview. They have all had to adapt to major changes in external circumstances, and have thus taken a further step in their evolution. Those who know or were part of the firms as they were during the period of the interviews will recognize that further changes have and are still taking place, which cannot be recorded here. No further formal investigations have been carried out, so such developments cannot be described and are not assumed in the discussion. The detailed stories, however, underline the constant process of evolution and change that takes place in the organizational structures and work patterns of all firms, so that any cut-off point represents an arbitrary moment at which to describe the present state of each organization.

Since the 1960s, partly as a result of the 1962 study which stressed the importance of such information,[1] the RIBA has collected statistics about the profession, based on its members, their numbers, the proportion employed in the public sector, the size of private firms, and the extent to which architects are involved in certifying building work. Apart from major growth – from nearly 21,000 to 32,000 registered architects between 1968 and 1991 – these statistics show that during this period the profession changed in structure less than might be expected. The number of private practices doubled to 6500, the result of growth in overall numbers and a decline in the proportion of architects employed in the public sector from 39 per cent in 1968 to 22 per cent in 1991. However, the breakdown by size of the private practices was very similar in 1991 to that in 1968 (Table 6.1). During the period, the majority (65–70 per cent) of firms employ 5 or fewer architectural staff, and the remainder are divided almost equally between firms employing 6–11 architectural staff, and all larger firms.

Table 6.1 Private firms in architecture, 1968 and 1991

	1991	1968
Architect members of the RIBA:	32,000	21,000
Percentage in private sector:	78%	61%
Percentage in public sector:	22%	39%
Percentage each size of private firm by no. of architectural staff:		
Very small 1–5	70%	65%
Small 6–11	15%	20%
Medium 11–30	} 15%	15%
Large 31–50		
Very Large 50+		

Sources:
RIBA Market Research Unit; RIBA Services Practice Register and RIBA Census of Private Architectural Practice.

The seven case study organizations each throw a different light on the context of practice and its many manifestations. If their experience is typical they present a picture of continual change and very varied adaptation. The consequences of change, and the need to be able to cope with it, are more vital for the larger firms, where changes tend to be more disruptive. The small ones form, rise, regroup, fall and rise again from the ashes of previous incarnations with relative ease. For the larger ones the management issues and tensions that changes bring to the fore require a set of skills and expertise that people in architectural firms are only gradually learning.

Choice of case study organizations

The firms chosen for detailed study are Roger Mears Architects, Stephenson Architecture, Ahrends Burton and Koralek (ABK), DEGW Group Ltd, Building Design Partnership (BDP), Hampshire County Architects Department and Pentagon Design and Construction Ltd. No attempt is made to suggest that the firms represent a statistically relevant sample, or a comprehensive range of possible types. There are some omissions, such as the study of a publicly quoted company, or an architects department in a private organization such as Sainsburys. Nor were they randomly chosen. They were selected by the authors on the basis of their knowledge of the firms, the respect that they command within the profession and the likelihood that their histories would shed some light on how changes identified in earlier chapters have taken place within architectural firms. These firms include archetypes that the profession has tended to assume represent a 'norm', and others that may be unusual or specially interesting to potential practitioners. They are there to offer vignettes of the profession against which to test the ideas that this book is exploring. The case study firms were selected to cover:

- *Different sizes*: Very small (Roger Mears), small (Stephenson Architecture), medium (ABK), large (DEGW), very large (BDP).
- *Different structures and ownership patterns*: Sole proprietor (Roger Mears), private partnership (ABK), limited company (DEGW), public authority (Hampshire County Architects Department).
- *Different services offered to clients*: Specialist service (Roger Mears), architectural design (ABK), design-based consultancy (DEGW), multi-disciplinary design (BDP), project management (Pentagon Design).

In addition they all demonstrate different design philosophies and working methods.

Study methods

The case studies were based on:

1. interviews with principals (and staff where relevant) about the firm as a whole;
2. records reviewed by a member of the firm's staff or the researcher;
3. selected projects (from records and interviews).

Material indicating the scope of inquiry was provided to the firms to prepare them for the type of information that was to be sought. This was in the form of simple questionnaires, indicating a minimum level of desirable factual material relating to the company and to any particular project examined (see Appendix IV). Interviews were conducted, based on the material initially provided, and records were reviewed where relevant and available. The object was to gain insight into the working of the practice rather than to focus on the design achievement. This involved looking for information about:

*   *Origins*: When, who, where, background of principals.

*   *Professional structure*: Numbers and mix of employees, relation with outside specialisms, with students and with new small firms, career structure and patterns for employees.

*   *Development and growth*: Changes in the above over time.

*   *Projects*: Size, building types, client types, sources of commissions and methods of obtaining them, numbers at one time.

*   *Management/organization type*: Degree of centralization of policy, work patterns and style of management of projects and budgets.

*   *Working methods*: Control systems in project design and subsequent stages, information systems, technical systems, publicity and marketing methods.

*   *Philosophy*: Type of staff sought and training given, type of clients sought and rejected and why, relationships developed within and outside the firm, as well as design methods and approaches.

Response to change

Earlier sections identified a number of issues which are to a considerable extent interrelated. In particular the issues discussed are concerned with the extent to which specialization has taken place, the ways in which new technologies have been absorbed, and the differences between large and small firms. The different case study firms illustrate a number of different responses to these.

Aspects of these are illustrated by the various cases. For example, the larger firms, BDP and DEGW, have adopted a wide range of specialisms, not necessarily the same as each other; and the viability of these within each firm, and therefore their importance, has fluctuated according to market conditions. The smaller firms have settled for far fewer specialisms, but, for example, the case of Mears shows that for a really tiny firm, as the majority of firms still are in the UK, to adopt a special skill and to aim to become known for this may be a good survival strategy. New construction technologies have played a part for those firms, such as ABK and BDP, concerned with advancing design ideas, and exploiting changes in these areas for the benefit of their clients, though none of the firms examined is a specialist in for example high-tech design. New technologies in servicing, seeking energy efficiency, have been an important concern for most design firms. ABK, for example, has pursued this field, in publications and in design (e.g. the Isle of White Hospital), while Hampshire has followed up issues of energy efficiency in schools, as is shown in one of the projects, Yateley School, discussed below. Of significance to all firms has been new technology. This has been relevant for drawing work, and computer-aided design (CAD) has been taken up by most of the firms, though the larger ones reached it some years sooner than the medium and smaller ones. Perhaps more significantly for office and project administration, without new technology even the smallest, Mears, believes it would be unable to compete. The particular history of Hampshire shows how, for at least one group of public sector architects, reductions in new building work created a shift towards maintenance and rehabilitation.

These case study organizations have all increasingly struggled with the need for management, marketing skills, and financial expertise towards the end of the period discussed. The need for training in these matters is felt acutely; the principals who responded to the questionnaire rarely considered it to have been adequate in their own cases. The tendency for the profession to consist of many small firms reinforces and is reinforced by the situation, both because the smaller firms have slightly less need of the missing skills, and because without these skills growth towards becoming a large firm is difficult to sustain. Without more training in how to manage a firm and adapt to the needs of their client body, rather than being educated in an atmosphere that prizes skills that impress their peer group, most architects may well be doomed to work in small practices and thereby be further marginalized in the building process.

Roger Mears Architects

A very small firm formed in 1980 in the wake of an earlier recession, Mears almost by chance has developed a specialism in conservation work. This is the feature that can distinguish the firm in the eyes of potential clients, from the many other very small architects' practices in London. It creates an identity that has survival value. Much of the work in the practice is however similar to other small firms, in being for small private clients. The management of the practice has been computerized so that Roger Mears himself is still able to run all aspects of it himself if necessary.

Figure 6.1: Roger Mears Architects: office interior.

Stephenson Architecture

Stephenson has had a varied career pattern in a number of different organizations, joining in partnership with several different groups while seeking a vehicle for his own approach to delivering a design service. He is now the head of the small Manchester-based practice which split off from an earlier partnership in 1988. He energetically pursues design issues in buildings and interior work, usually for commercial clients, although they have done some private sector housing. Staff are included in the decision-making process in the firm, though the processes by which this happens are suited to small groups, and would need redefinition if the practice were to grow substantially.

Having sought a pattern of work where management and hierarchy are subordinated to design effort, Stephenson has created a situation where the firm is dependent on him to take the overview and ensure that the direction in which it is heading is viable. This leaves him personally less time working on designs than he would originally have thought was ideal.

Ahrends Burton and Koralek

ABK can be thought of as almost an archetype of the traditional architectural practice. It is medium sized, and though its size has fluctuated continuously, according to economic conditions and luck, it has never strayed far over the boundary to having more than 30 architects. The intention of the three partners, friends since college days and able to start the practice because of a competition win by one of them, has been to maintain the practice as one where they can keep all projects under their direct personal control. This creates a natural limit to expansion.

Figure 6.2: Computers are now an integral part of ABK, both in design and production, and in the various systems required in practice.

Another corollary of this fact is that very large projects have a considerable impact on the size and management of the practice. Despite avoiding the stresses of extensive growth, the firm has found, like others, that it has become important to tighten up management habits, acquire financial expertise, and develop systems to handle the ever more complex communication requirements of modern offices. The work done by the practice is entirely architectural. They do not employ professionals from other disciplines although they have regular working relations with some from other organizations. They are committed to exploring innovative design ideas in as wide a range of contexts, both project type and construction method, as possible. This means that they are generalists rather than specialists in any one type or approach. They design both buildings and in some instances urban areas. Recent projects have seen the beginnings of a move to providing the design skills for a broader team where they may no longer take the role of team leader. This is not a sought-after objective, but has proved to be a workable relationship.

The staff that join them are selected for design ability; the 'good all-round architect', who 'is sensitive to design questions and understands construction' and so can work on detailed design. As the firm is unlikely to grow, the positions of partners and associates being effectively filled, there is some dispersal as young people move on to other firms or to start on their own. The relationship of ABK to those who move on is usually supportive, so that in reflection of their own experiences when they started out, they will, at least in good times, pass on work that they are unable to take on themselves.

DEGW Group

This multi-disciplinary practice is large by the RIBA's standards, having for long periods had more than 30 architects employed in its various offices. They were chosen as a case study because they have distinctive differences from many other firms, and represent an area of specialist expansion for the profession. The London office is the largest, and most of the others are overseas. DEGW was established early in the study period. It offers a range of services and consultancy not traditionally thought of as the mainstream of architecture. For its age and size it has a smaller than usual repertoire of new buildings to its name, and a larger one of specialist services. It has concentrated its services in areas that do not depend on new building work, such as interior space planning and strategic appraisal of buildings, brief writing, and strategic facilities planning. It has thereby created and partially captured a niche in the world of buildings. Specialized teams have been built up in the different areas of the practice.

The firm's most consistent objective has been to consider buildings from a user's point of view before an aesthetic one, generally concentrating on commercial buildings, where the user is not only the individual working in the building but also the wider organizational entity that the building must serve. At all times there has been a respect for systematic collection and use of relevant factual information, as well as use of design skills to solve problems.

■

Porters North Handbook

Welcome

The move to Porters North is a watershed. It is very important for DEGW, not just for the London business, but for the Group as a whole. As well as being the platform from which we are going to launch a new company, it is a symbol of our determination to give ourselves the best possible opportunity to further improve the services we give our clients and thus ultimately ensure our own success.

A great deal of financial and emotional investment has been made by many people in selecting the location and planning and designing the facility, as well as devising the technology and support infrastructure. I'm sure that not only will it pay off for ourselves, but will be seen to set new standards in refurbishing and fitting out these kinds of buildings.

This handbook is to help familiarise us with King's Cross, Porters North and its facilities. I hope you will enjoy reading the Handbook, feel informed about the location and the operation of Porters North. Above all, I hope you enjoy working in the new building.

Colin Cave, managing partner

DEGW

Figure 6.3: DEGW practice what they preach and provided guidance on the new office when they moved in 1989.

With a special interest in and awareness of the importance of the nature of organizations, DEGW has made a continual and self conscious effort to create an appropriate organizational structure for itself. This has undergone a series of modifications, considerably influenced by the change from a partnership to a limited company which took place in 1989. The need for much more expertise in management, both of finances and of the company as a whole, became apparent at this time.

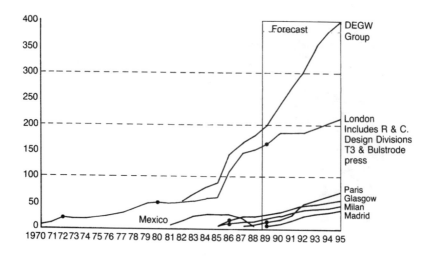

Figure 6.4: DEGW Group: an office move every 8 years or so to accommodate growth. In 1988 past growth was revived and predictions made. The subsequent recession altered the pattern.

This need was the greater because the firm has always included a galaxy of smaller alliances and concerns, reflecting the directors' interests in worlds alongside that of architectural design, such as furniture design, information technology, graphics and publishing. The written word has always been of extreme significance to this practice, whose reputation has been consolidated through articles, books, and published research reports. They are selling an approach, including teachable skills, not a style which can be copied, nor innate creative ability.

Building Design Partnership

This is the oldest and largest of the practices that we studied, having a 55-year history since its earliest founding by Grenfell Baines. Age and size are not necessarily combined. Practices of many years standing exist which have chosen or have happened never to grow, while others grow to a large size in a very short time. None the less BDP is unique in this country, having employed over 1000 staff for a number of years, more in the style of large companies in the USA. The firm is made up of a number of individual offices, which somewhat reduces the apparent size.

A detailed history of BDP provides not only an interesting record of changes in design development but also, by virtue of the range of disciplines and locations, a picture of the context of architectural practice in the UK over

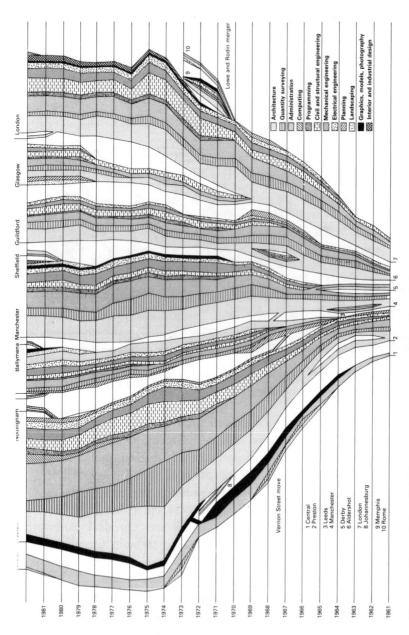

Figure 6.5: The evolution of BDP between 1961 and 1981 shows the range of activities varying over time and by office. Systematic record keeping of numbers of staff, structure of offices, and breakdown of output has characterized BDP since the early days.

101

the last half century. The firm has remained faithful to the original founder's beliefs about the organization of design, that is that it should be on a fully equal multi-disciplinary basis, and that the staff should have a share in the profitability of the firm and therefore a stake in its performance. At different times and places the range of disciplines and their relative importance and success have varied, and there is a constant debate about the extent to which some of the minor disciplines should be a central specialist resource available to all projects anywhere in the various offices; but the major groups, architects, engineers of all sorts and quantity surveyors, have tended to be prominent in all the offices.

Figure 6.6: The London office of BDP. The centre of gravity of the firm has been gradually moving away from its original in Preston.

The firm is not like an assembly of smaller firms each headed by a well known prima donna – not a summation of Rogers, Foster, Stirling and others. There are of course a large number of internationally renowned individuals with their own specialist skills or knowledge within the firm. However BDP does not set out to enable a particular design approach to dominate, but seeks to produce effective buildings in many different styles and idioms. The firm has demonstrated an increased willingness and ability to provide a partial service, putting in some parts of a project but not the whole multi-disciplinary team that it would be able to offer, and to work with designers and different team members, in the combination that best meets the clients' needs.

The firm has developed a number of well thought out management systems, and has kept a consistent set of records and information about performance over many years. This considerable management effort is thought not to dominate the professional and design objectives which are regularly reassessed and communicated to the staff. There is a well defined concept of career progression for individuals.

Hampshire County Architects Department

As a local authority architects department, Hampshire represents a form of practice which has been particularly vulnerable to the economic and political changes of the last twenty years. The department was radically transformed with the appointment of Colin Stansfield Smith as Chief Architect in 1974. The management and design approach were changed, by introducing new staff, and by altering the structure from one based on building type to one based on geography. The department has gained a wide reputation for exceptional design quality, especially in the building of schools, a major part of their output. In the early days of the new regime outside consultant architects were sometimes commissioned, allowing the quality of design to be seen to be of vital importance. By 1991 the situation had been reversed, Hampshire's reputation being such that they could qualify as the high-flying outsider that 'could perhaps be called in by another organization for design skills and expertise'. The workload was originally entirely dependent on the local authority as client, though the department is now more able to look elsewhere. There has been a change in the types of buildings and the mix between new build and refurbishment as the requirements of the local authority have changed. The department has also a large number of professional staff, engineers, surveyors and others, whose function is to maintain and manage the building stock owned by the council, a formidable number of schools, fire and police stations and public buildings. The facilities management skills that could have been developed here have not formed part of the thrust of the department's sales pitch for their skills. This is the opposite phenomenon from some other local authority departments. Hampshire have not sought to establish themselves as a design group with a special focus, except as designers, despite their expertise on schools and on energy use in these buildings. This may make it appear that they are joining the race with well established private firms somewhat late. It will be interesting to see how they progress in the next few years.

Pentagon Design and Construction

This is a design–build company, not an architectural firm. Its role is that of linking a client and his needs with the means, both in design and construction, to achieve them. Some might think it represents the newly found opposition to traditional forms of practice. Pentagon developed as an organization in 1985, from an earlier experience within a design–build arm of a construction and product sales company, ready to exploit the commercial building boom, and with a unique selling proposition, the maximum guaranteed price. Architects who work with them speak well of their contribution to the building process. Headed by a multi-disciplinary group of directors, especially strong on finance and engineering, they subcontract the major part of design work and are not the employers for the large teams needed to produce large buildings. This gives an extremely robust way of dealing with fluctuations in demand.

Projects are not carried out in the same sequence as many undertaken by more traditional firms, although architects who work as design–build project managers as well as in more traditional ways will recognize the process. A development plan is put together, for example, by the client and Pentagon, and perhaps with the design firm. Pentagon will suggest a price, at the concept stage, based on cost per unit area ($£/m^2$) and related to the amount of space, established in discussions with the client, that the project is likely to gain planning permission. If this price is suitable then a scheme is developed to meet the site planning requirements and the client's brief. The design team who develop the scheme will include a firm of architects, specialist engineers and others as required, generally chosen by Pentagon. Major subcontract packages are incorporated, rather than seeking the services of a single contractor, which is a more common form of design–build. Traditional design to stages C or D of the 'RIBA Plan of Work' will be carried out, and a bill of quantities prepared. Tenders for the various subcontracting packages will then be sought. In the case of the services a detailed performance specification may be prepared by the engineering contractors on the team, and this will define what the job requires but not how it will be distributed. A number of design and install subcontractors will then compete to provide a firm price. Prices are needed rapidly as the maximum guaranteed price is given to the client prior to being asked to go ahead with the project. In this process Pentagon are deciding who will design which parts of the project, and setting the boundaries of the budget.

Changes in practice

The RIBA's 1992 strategic overview of the profession indicates a series of changes that have taken place in practice,[2] and all of these were reflected in the experiences of one or more of the case study firms.

The 1960s was an optimistic decade, when flower power seemed a real possibility and architects were generally positive and enthusiastic about their ability to have a lastingly good effect on peoples lives.[3] In the early 1990s, even without the current deep recession, this mood of optimism would seem almost naïve.

Clients are seen as having become more varied, often more sophisticated and better informed, both about what they want from their project, and about what they expect from the construction industry. Some of the buildings that they require have become more complicated, and represent a greater financial commitment, so they are more demanding, especially in relation to cost. There is a greater influence from private sector clients, with their strong interest in balancing time and cost as well as quality. This can sometimes take the form of an unrealistic expectation of the speed that can be achieved, to the detriment of the design if insufficient time is allowed. The impact of these changes was expressed in a general realization that more documentation is demanded by clients. They are likely to expect to play a greater role in management, or to have representatives of their own with management skills, particularly for cost control. The belief that creative energy is being diverted into managing increased documentation was both expressed and resented by several of the firms, particularly those whose service philosophy stressed the importance of design. The real need to provide an alternative way of managing the documentation, which does not demand energy from inappropriate sources, has not apparently been fully explored by these same firms.

Clients are not so likely to follow a simple appointment procedure such as taking soundings and choosing an architect on the basis of trusted advice. They are now more likely to run a complex selection process comparing a range of possible design teams. Fee bids have become an important feature, which did not exist before, and can now take 10–20 per cent of a senior staff member's time. The relationship between client and architect is far less one of gentlemanly trust than it used to be, and one informant commented that as soon as you are no longer treated as a gentleman to be trusted, you cease to behave like one.

All the case study firms were concerned with quality assurance, as one expression of their recognition that clients demand some proof of ability to manage and deliver a product, as well as to design. A further subtle change relates to public clients. Following the demise of the government Property Services Agency (PSA) public clients have lost most of their knowledge of building procurement and have therefore become much more like one-off commercial clients who may have little or no experience of getting any project built, especially a complex one.

The procurement process has also changed. Instead of a single fairly standardized procedure, there is a much wider range of contract types, and an increase in procedures that appear to threaten the role of architects, namely design–build and project management. The range of contract types is particularly well understood by the larger firms, such as BDP, with its large range of building types and clients. For larger projects there is likely to be the loss of leadership to a project manager from another team, usually from another discipline, often a quantity surveyor. To a small degree architects have themselves responded by specifically offering project management as a skill, but there is little evidence that clients are particularly impressed by this offer. Clients often prefer cost or management specialists for a role that they do not see included in the design process, and have little reason to believe that architects do particularly well. Architects perceive the need to recapture their project manager role. They have a difficulty in refocusing on what the architect's role should be if 'team leader' or 'project manager', is not a significant part of it. Architects need to be educated, or retrained, to recognize from the start in each situation what their role is or should be, and how to perform it, rather than to expect to be the team leader in every case.

The building industry is less craft based and does not consistently offer the same level of skills as in the past, so that where architects have high levels of expertise in construction matters they can make a more vital contribution than before. This was particularly mentioned by ABK. For the smaller simpler projects, the contractor could once have been expected to be able to have the required expertise among his own staff. This is not always the case today. Buildings have become more complicated; components are often factory made; site assembly may be done by casual and unskilled labour. When a management contractor is in charge the direction of the project may lack practical building expertise, so there is more dependence on the architect getting the details clear and correct. The architect 'is becoming the building expert', replacing the contractor's expertise. There are fewer experienced builders; a 'contractor' is not a 'builder'. Contractors ask for more and more 'clarification' even down to trivia such as 'what nails shall I use'.

Strategic choices

In developing over this period, and responding to change, each of the case study firms has followed different routes, sometimes from choice, often less deliberately.

Size

Whether to grow if opportunity offers, how much, and in what way have been differently approached by several of the cases. A strong urge for some has been to retain control of the design process in the hands of a few original partners. This has a survival value, in that the natural limit this places on growth means that the firm is likely to be resilient over the cyclical booms and recessions typical of our economy. At the same time, it avoids the need to adapt the organization to cope with increasing numbers.

Radical reorganizations seems to be a particular necessity once a group reaches around 50 people and is still growing. Both DEGW and BDP, the larger cases, have put large amounts of energy and thought into frequent restructuring. They have sought forms of organization that will control and utilize the expanding population and specializations of the firm, and to create structures that could outlive the original founders. BDP has already succeeded in this latter aim.

Organizational restructuring is also undertaken because of the need for internal reappraisal. Decisions about future directions, and responses to external pressures, seem to be as influenced by the desired relationships with the outside world as by the actual situation outside. There is an inherent dilemma in growth. Often the first faithful supporters and workers, who are vital to help a small firm become established, have not got the necessary skills or experience to play their original roles in a large one. The difficulty in reconciling this fact with loyalty to the first people on whom the firm originally depended utterly may be why so many firms stay small to avoid the problem, and why the larger ones have so many attempts at restructuring to handle increases in size in the least painful way.

Technology

The firms have all gradually introduced CAD. This has been an inexorable development, but in fact the cases studied did not seem to embrace CAD enthusiastically, despite recognizing its advantages, and accepting that clients frequently expect it. The age of the partners involved, and the fact that they were not trained in the use of CAD as part of their education, may have something to do with this. They consider that the mechanical aspects are a constraint upon some part of their identity as designers. The demands of technology have generally been relegated to younger, more junior members of staff.

Management skills

The increased complexity of the context in which architects work has increased their need for management skills. Most of the firms had taken on specialized financial management, usually rather recently, partly in response to the increasing need for special financial understanding created by the introduction of VAT, the increase in indemnity insurance premiums, and other similar

107

developments. Only the design–build firm had a financial manager on board at the very start. More telling is the fact that most of the firms found their first efforts at delegating financial management to a specialist were not successful. This seems to be evidence of a lack of clear understanding of what to expect from this role, and how to be sure that it is being delivered. This is a clear gap in the architects' armoury which needs to be filled by education and by a change in the general attitude towards management. All forms of management were considered of less interest than the creative act of design by a significant number of the senior staff in the case study firms. This is in part a consequence of the predominance of small firms, where absence of management, while to be regretted, is not often fatal. Senior staff in small firms have relatively little experience in really large organizations, of which there are so few in the field, and then not usually in positions where the management processes are truly apparent; time sheet keeping may be their only contact with this side of the business, where their early experience may be gained before they set off to start up on their own. This cannot augur well for a profession that wishes to retain, or regain, the project management role that it seems to have lost.

Marketing and specialization

The attention given to the origins of the case study firms is partially to indicate the great importance of early influences on the way in which people come together and work together. It is also to indicate that the intricate web of connections that lies behind most of the firms' client bases starts at an early stage. None the less, there is an increasing need for firms to market their skills in order to capture a sufficient share of the changing client population. Interestingly, the most common form of marketing is still to describe the projects completed in technical journals and books intended to be read by fellow professionals. While this reinforces the profession from within, it does not necessarily offer a clear route to the client.

DEGW was chosen as a case that showed how specialization could find a niche market. In the course of the period studied, there has been a great increase in the number of companies offering a similar service, both management consultancies and other architectural and design organizations. The competition has grown harder. At the same time the market for these services has also expanded, as clients have come to understand the possibilities and benefits that are different from those that they have been accustomed to seek from architects generally. This is partly due to the increased activity by multinationals, since they homogenize a market to some degree, thus emphasizing the importance of ideas from the USA, other parts of Europe and now Japan. None the less, it can be seen that though specialisation may help to carve out a market, once the specialization is generally recognized, there is a need to remarket, to retain and expand what may only recently have been an exclusive client base.

What to market, as well as how to do this, are not clearly understood by all firms. The RIBA *Strategic Study of the Profession* emphasizes that what architects are seen as doing best by those outside the profession, is exclusively 'design', and suggests that therefore an effective marketing strategy for the profession as a whole is to reinforce, to the general public and to clients in particular, the great importance of the value of design. This, once clearly defined, may be a route that will produce a better climate for the profession at a generalist level. It does not tackle the problems of a specific firm.

In some cases, design capacity may be increased by association with another firm of architects. But other questions arise about how to package the design service if it does not include team leadership; how to relate to other design groups when the timing and level of their involvement is controlled by others; how to decide which specialisms to offer, and whether to consider these as separate disciplines or as an integrated set; whether project management can be recaptured in particular projects, and if so how to do this if management is in the hands of accountants or quantity surveyors. This complex set of questions to be addressed and somehow answered by each firm makes the task of deciding what and how to market much harder. Talking about design is easier. But it is unlikely that the answer to what and how to market lies in educating the client/public to appreciate design, in order to be able to market this elusive product. Better information about what clients and the public as a whole have wanted, and what they have found satisfied their needs, will help firms to make relevant decisions about where to position themselves in the field and how to market their services. A clearer knowledge of the building construction industry, and its role, methods and capabilities, and the part played in this by architects, will put into perspective what it is that architects can do, and do well and profitably. Looking outside the profession, to the general context rather than talking to each other about the subject closest to their hearts, namely design, will be vital to future choices.

Staffing strategies

The choices that firms must make, as they adapt to their context, all affect the type of staff that they will need to employ, and the ways in which job satisfaction can be achieved for these staff. Different approaches have been adopted in the cases: offering a full integrated in-house service covering all areas of design (BDP); adopting differentiated specialisms (Mears, DEGW), taking on teams to deal with each project as it arises (Pentagon); and holding a core of committed staff, with all additional requirements staffed on an as-needed

109

basis (ABK). These approaches will have different staffing needs, different training and different management and marketing implications.

Working for public clients

There have been more dramatic changes for public authorities than even for private firms in this period, as the public role in provision has swung from rebuilding after the war to a reduction in overall need for new building, and to a far more market-oriented approach, less accountable on the basis of social need, and more financially accountable. Schools building in particular – and Hampshire is a good example of this trend – has slowed down, and the stock needs to be judiciously culled rather than expanded. Local government activity is now greatly reduced, with authorities having fewer powers and a smaller need to fill, so architecture departments around the country are all having to find work elsewhere. Financial constraints are more apparent, accountability is demanded, and financial management techniques have to be in place and demonstrated. These changes add up to a completely altered micro-climate in which the plant, a local authority architects department, is now expected to survive and flourish, even if not to grow in size. There is the possibility that as activity slows to a halt, where public authority architectural departments are disbanded or reduced to maintenance management divisions, the knowledge base will be eroded, the links between briefing and procurement will be broken, leaving fewer people in the public departments with knowledge of the role and structure of briefing. If and when new buildings are required these skills will then need to be rebuilt. Luckily in Hampshire this is not yet a problem, but the effects of the demise of the PSA with its previous central role in procurement of government buildings can already be seen to indicate potential future problems.

A view of the profession

The period during which this research has taken place has demonstrated vividly that architecture is constantly in a state of flux and change. We have only studied seven cases, but they have covered a wide range of philosophies, sizes and market positions, and they have all had to respond fluidly and rapidly to the changes they have experienced. These have taken place in their market, in the skills they can call on, and in the aspirations they can possess. The cases provide a basis for some of the generalizations that have been made concerning aspects of the future of the architectural profession. From these cases it seems that the profession is both varied and flexible, and a unitary view of the future of the profession may be too limiting for the potential that is inherent in architecture and building design today. Attempts to create a single framework by which to understand what is meant by architectural practice can help to focus ideas in the face of a changing market, but can also lead to the possible danger of inhibiting a flexible response to currently unsuspected opportunities. This observation may be valuable for educators as well as for regulators of the profession, and for those who collaborate with architects in the construction industry. The following two chapters give an abbreviated text of the case studies, described under a common set of headings.[4] There are two groups: the first consisting of the more traditional types of practice, the second of those with unusual or novel approaches. The lessons and strategies discussed above can be seen to have affected all the organizations in different ways for different reasons.

Notes

1. *The Architect and His Office*, London, RIBA, 1962.

2. *Strategic Study of the Profession*. Phase 1, *Strategic Overview*, London, RIBA, 1992, p. 21, table.

3. 'Perhaps [the architectural profession's] most serious problem is how to achieve, without lowering its standards, the great increase in its output that is needed to cope with the demands of the building boom – demands which are expected to go on growing faster than architectural man-power for as long as peace and prosperity may last. For the forward looking architect these are indeed exhilarating times.' Senior, Derek *Your Architect*, Hodder with RIBA, 1964.

4. The full version was published as Eley and Symes *Recent Changes in Architectural Practice*.

7

Developing the traditions

Roger Mears Architects

Description in 1991

Since setting up in practice in 1980, Roger Mears has run an office with one to five architects including himself. His practice has therefore always fallen into the smallest bands described by the RIBA census, that is one–two and three–five.[1] The office generally carries out work to existing houses, either repair and renovation, or extensions and additions, and is run from Roger Mears's own home in Islington.

Origins

Roger qualified from Cambridge and had no initial ambitions to run a practice of his own. His early experience was with housing. A large project on which he was job architect, some housing for Camden, led to an opportunity to do complex working drawings and establish over time an excellent working relationship with the contractor. Working for Castle Park Hook and Partners on a housing scheme in Haringey, he became the site architect in a small shopfront office of which he was in charge. Without consciously setting out to do so, he began to discover what it would be like to run a small office on his own. When the 1980 recession started to bite, his employers were forced to cut back, and Roger found himself unemployed and ready to start as an independent practice.

Development

Location
The office started in his own home, in an upstairs room. His own typing skills were good, and he realized that he could do all the secretarial work that was required on his own. However he did eventually need to employ architectural

assistance. By 1986 the office complement was four, too many for his garden studio, and a small three-storey, three-room shop/office was rented around the corner. A search for a larger family home was extended to look for a house large enough for both the family and the office; they all moved there in 1988.

Practice activities

The work of the practice has been with domestic buildings, for private owners. While this is not an exclusive policy, it is a self-reinforcing situation, since most clients are found through recommendation. The range of work includes repairs, for example to roofs, as well as modest and more ambitious extensions, particularly conservatories. The largest jobs in the practice have had a particular emphasis on the restoration of historic and listed buildings.

Size and turnover

The practice consists of five people, all architects, and has never been larger. For several years it was four. The largest project in the office is for nearly £2 million, but this is not the staple diet of the practice, despite the fact that it has been of huge significance. Generally projects are much smaller: £80,000 would be a reasonable contract size. An average of four to ten such projects are completed in the course of a year. As many as six on site at one time is above average and not desirable; two or three is more usual.

Projects and clients

The large majority of the early work of the practice was on old buildings, usually houses. Working on a Welsh water-mill conversion Roger became interested in the origins and construction of old mills, and joined the Society for the Protection of Ancient Buildings (SPAB). This eventually led him to take a course on old building repair. The specialist skills that the practice has developed now combine an interest in the domestic requirements of individual owner-clients with a knowledge of how to bring old buildings back to the state they were designed to be in. The projects that the practice carries out and the clients it works for fit into this framework of skills. Most clients are found through personal recommendation from other satisfied clients, and there is some repeat work.

Practice behaviour

Company structure

The group of five can hardly be said to need a major structure, and what there is is informal. Roger himself is in full charge, the others are employed by him, and they meet once a week in an office lunch to catch up on the overall work progress and discuss important issues.

Staff selection and career patterns

Roger has selected staff particularly able to handle the historic building work, and sympathetic to his design approach. Generally he has found staff through personal contacts, and takes great care to ensure that they are in tune with his design style as well as the nature of the practice.

Management issues

His experience with Castle Park Hook, his first employers, has led Roger Mears to do a lot of his standardized work on a computer, using particularly word processing and spreadsheets. This has meant that he has been able to manage the administration aspects of the practice without assistance, but it has had the corresponding disadvantage that he is able to manage these tasks far more efficiently than any newcomer. This means that he ends up doing a large part of them himself; his staff never really learn to exploit the systems to the full. Generally each job is sufficiently distinct that it is not possible to use standardized specifications, despite the fact that it was for this use that he first gained computer experience. Notwithstanding this comparatively early introduction to computers, he does not use CAD as he has not so far found a system that he feels is suited to his work requirements.

Projects are now managed by a member of the office with a reasonable degree of independent responsibility. While Roger expects to be made aware of any difficult issues and to participate in the decisions that these may need, he does not manage all the projects. Because of his experience he still works more economically than the others on the smaller projects, so he ends up running many of these himself.

With three other small architectural firms in the same part of London, Roger Mears has been investigating the implications of quality assurance (QA). Without expecting to attempt to become registered in the immediate future, he is interested in what the practice can learn from investigating QA, and from discussions with other small organizations like his, about management and practice behaviour.

Financial management

As with other aspects of administration, Roger handles the financial manage-ment of the practice himself. Projects are not tracked individually against their budgets, but the overall financial position is reviewed on a quarterly basis. The staff keep time sheets and so does Roger for some projects, though he does not do so for any activities not charged on a time basis.

Marketing

There is a practice brochure now in the office, two or three years in the gestation, and not yet systematically dispatched. Jobs have mainly come from direct personal recommendation. The booklet provided by his local authority, describing for members of the public how to apply for planning permission, will also soon carry an advertisement for the practice.

Other associations and roles

There are no other professionals formally related to the practice, but the firm generally uses the same quantity surveyors (QSs) and structural engineers. The QSs that they are using on the largest of their projects, Tudor House, has special skills in relation to historic building preservation work.

Philosophy

When he started on his own, Roger had few illusions that he would imprint his artistic concepts on others. He feels that for a client's private home, work done using the client's own money should be done to the client's personal satisfaction. If he wants a particular effect, or doesn't want to spend money on some specific work, then that is his choice. At the same time it is clear that if a prospective client is not likely to be prepared to pay for work to the level of finish and detail that Roger expects to carry out, then there is likely to be a mutual parting of the ways, hopefully before work on the project has even begun. The SPAB's philosophy of 'conservative repair' informs all the work of the practice, and close attention to detail is sought in all projects, as well as the use of good materials that will complement existing buildings.

Roger expects to be involved in the design of all projects in the office; it is his office. However in principle he believes that it would be possible to come across someone with whom he would be able to work in such accord that he would enter into some form of work partnership. This hasn't happened yet.

Stephenson Architecture

Description in 1991

Stephenson/Mills was established in Manchester twelve years before our study. With twenty architects and technicians, and six interior designers, they were chosen as a case study of a medium-sized practice, firmly based outside London, with a growing workload in a range of building types, and a growing reputation. Many of their current and recent projects, in offices, college buildings and museums, are within a few minutes walk of their central Manchester office.

Origins

Though Stephenson Architecture was only formed in 1989 (and has since changed again), Roger Stephenson qualified in 1969, at the start of our study period. His route to Stephenson Architecture reflects the amoebic splitting and regrouping that is common in developing architectural practices and will have a familiar ring to many other architects.

His first job, after qualifying from Liverpool, was with BDP in Manchester. After three years there, he joined Michael Hyde, who started a Manchester office for Dry Halasz and Dixon. In 1975, they split from the London parent to become Michael Hyde and Associates, where Roger Stephenson was one of four partners. Despite common ground at the start, Stephenson eventually decided to practise on his own, taking no projects from Michael Hyde but following his own set of leads, outside the mainstream of Michael Hyde and Associates. They bore fruit, and he needed to employ a student after his first two months. Eventually, after seeking hard to find the appropriate person with a commitment to and skill in design, he was joined in 1981 by George Mills. In 1983 Stephenson/Mills was formed, which flourished; but in 1988 it divided, splitting the staff, the client list and the workload, to form two separate practices, Stephenson Architecture and Mills Beaumont Leavy.

At that moment Stephenson Architecture consisted of sixteen people. A new structure was created including a salaried partner and four associates. The partner, Philip O'Dwyer, and one of the new associates, Jeffrey Bell, both had contacts with Stephenson that went back a long way.

Development

The following description of the development of the firm, covers the period since 1979 when Roger Stephenson started up alone, and treats events that occurred during the shifts to become Stephenson/Mills and later to dissolve it as part of a continuum of a single practice. (More recent changes, since 1992, are not considered in this description.)

Location

The office premises has at all times been both symbolic and vital to Stephenson. The first office, in King Street, was chosen to be central for the 'credibility' that it afforded to the one-man band. The next, vitally significant office, was spotted by Stephenson in 1983 in Castlefields Conservation Area. St Matthew's Methodist Sunday School building, said to have been designed by Charles Barry in 1827, was eventually acquired and turned from a decaying wreck into a prize-winning office. Its layout was designed to foster the interrelationships and communication that Stephenson was committed to in his concept of an architectural office. In this it was only partially successful. Its illustration on the front of the *RIBA Interiors* supplement[2] was instrumental in ensuring that they were commissioned by Siemens on their largest office project at the time. After the split of Stephenson/Mills, St Matthew's was sold,[3] disciplines or again performing an important service – providing the equity for independent survival. The new Stephenson Architecture again invested time, effort and capital in its premises. It was vital to have its 'own front door', to create a space that could display its design approach and capabilities, and to plan space that would allow the right social relations to take place within the office. Its central location in Crown Square emphasizes the strong local base that the office has built up, with its short distances from so many significant areas of Manchester where the practice's work is located.

Practice activities

Stephenson was keen to diversify the workload of the practice. It did start with about a third of its work in housing, largely based on old contacts. This steadily dropped to 10 per cent by 1985–8 and has currently disappeared. It has been replaced by a steadily growing office component to the workload, now about 40–50 per cent with some industrial space nearly always on the books.

In addition the description of the firm has always included the term 'interior designers', in order to attract refurbishment and interior work as well as the design of new buildings. In the early days, the end of the 1970s, 'there was a great deal more refurbishment work about than there was new build', as Stephenson said in 1987.[4] The interior design work started out by being mainly for office interior refurbishment, but has really grown (fee income amounting to 30–40 per cent) with the work for catering areas in the terminal buildings at Manchester and Heathrow airports. There has also been a small regular amount of work that falls into the town planning and feasibility study bracket.

Size and turnover

For the first few years the office was essentially one or two architects and a secretary. From 1983 it experienced steady growth till the split in 1988, as the flow of projects gathered momentum, and the practice became established. It was of course dramatically affected by the split. In 1989 it was almost down to half its previous size, but by the end of 1990 it had doubled, and was larger than ever. Senior members of the practice profess to aim to stay small, and to be willing to restrict the workload, should they be so lucky as to have a choice, in order to ensure that the firm does not grow too large, fearing that larger than the present 32 staff would cause problems.

Projects and clients

Examining how projects have been commissioned shows a distinct pattern of small projects begetting much larger ones. For example, in an early new-build office project, a 1800m^2 block in suburban Wilmslow, the developer had been impressed by a house extension done by the practice. The smaller office projects and the office interiors work have acted as a foundation for further office work. The newest work that is coming into the practice includes a project where the client needs their skills at dealing with listed buildings and the procedures for consents, which have been demonstrated in the intriguing solution devised for the John Dalton Street project.

Friendly personal relations between senior staff and clients have been of great importance. The related issue of how large a practice can get before the personal involvement with the details of particular solutions is lost, and what changes take place in the nature of the practice as shifts in this relationship occur, is one with which the practice is grappling, in their deliberations on size and on company structure.

Practice behaviour

Company structure

Stephenson left BDP to get away from some of the disadvantages that he perceived in a strong vertical hierarchy, where career advancement led away from design and towards administration. So one of the basic beliefs that he brought to his new firm in 1979 was that all principals should always have active design projects. The practice description that is included with current project proposals states that they 'keep internal hierarchy to a minimum to enable senior staff (including partners) to pass on the benefit of their experience by continuing to practise, avoiding as much as possible the drift into non-architectural management and administration'. The way in which this aim is embodied in the current firm means that the two partners and four associates see themselves as sharing the load of design direction and of practice management. Each of them is responsible for some administrative aspect of the firm. At the same time, with the exception of Roger Stephenson himself, they are each in charge of live projects. Stephenson felt after the 1988 split that he would have to concentrate for a while on ensuring continuity for the practice, and has since discovered that he can share his design concepts more effectively by taking a 'tutorial' role with respect to any and all projects without 'owning' any of them.

The overall aim is a reduced hierarchy, which has been embodied in regular shared discussions about where the practice should be heading and how it should get there. This has been one of the attractions for the present staff compared with the Stephenson/Mills structure. The firm is at a crucial stage where various structures may be adopted, including continuing as they are. More than one may have to be tried before a final one is agreed by all concerned.

Staff selection and career patterns

Quoting again from the project proposal document, the aim is 'to maintain a "design driven" group of talented professionals devoted to the practice of architecture and all aspects of design'. Now they also expect to look for organizational skills, for designers with their feet on the ground as well as with an artistic ability and enough flexibility to be innovative. The current senior staff are in a special situation, as with Stephenson they have all participated in the process of starting up a practice in very favourable circumstances. This has greatly helped in feeling able to share all the management roles and minimize hierarchy. There has so far been little turnover amongst the architects.

There are six technicians, a relatively high number, who are employed at lower rates than architects. To get the right level of technician is difficult, as those that are older and more experienced may not be imaginative enough to contribute to the design process, but the dynamic ones tend to move on, often training as architects, as Philip O'Dwyer himself has done.

Management issues

The practice management is handled in meetings to discuss workload, which take place every two to four weeks, and meetings for all the partners and associates to examine practice objectives. The latter have been at six-monthly intervals but currently are happening more like quarterly, as Stephenson has felt it important to ensure that all the senior staff are more deeply involved in these aspects of the practice.

An interesting facet of how projects are managed in this practice relates to the concern that large project teams should not become isolated, allowing factions to develop in the office. This stems from the belief that the split between Stephenson and Mills could in part be attributed to the fact that Mills's work was dominated by the large and demanding Siemens project, for which a team of six to eight people had been put together. The largest project in this practice, John Dalton Street, has needed a team of nine or ten people for long periods. This group has deliberately never all been brought together in one place in the building. The disadvantage of needing to walk about to meet up with team members, and thus being away from one's own workplace, is seen as having the greater benefit of ensuring that people see and interact with others in the office with whom they may not be currently working.

Financial management

It was realized by 1987 that some trained financial assistance was needed to take some of that side of the administration over from the senior designers. In addition when the split from Mills occurred, one of the decisions of the practice was to invest in a computerized accounting system. The first appointment of a financial manager was not a success, but the office administration and accounting systems are now in the hands of someone who, though initially not experienced in the needs of architectural practices, had experience of the management role in other businesses, and the management is increasingly professional.

Marketing

Much of the early marketing took the form of intensively making personal contacts at every opportunity, for instance by attending professional functions. The firm now also employs a London-based publicity agent, specializing in architectural practices. Her role effectively comes down to linking the practice with opportunities to publish their building projects in magazines. The other way in which they spread information about themselves is in the elaborate A3 documents that accompany project proposals, and the practice has begun to concern itself with how to reproduce these effectively to aid publicity.

Other associations and roles

The firm has a strong interior design component and wider planning is done when relevant to building projects. However the practice predominantly provides an architectural service and calls in the services of other consultants as required, rather than employing their own staff with separate skills. When first on his own, Stephenson taught one day per week at the Manchester Polytechnic. This connection brought him his first student employee, now an associate, Jeffrey Bell, and the practice is now in regular touch with Manchester University, contributing to design studios and crits.

Philosophy

Not only Roger Stephenson but other senior staff are clear that they are in practice because design quality matters. There is a belief that the size of the practice has direct relevance to this, and an unspoken belief is that good design must come from the senior staff. So there must not be too much work for those people to have both a hands-on contribution to design, and also a capability to control quality in the project teams. This is in effect a statement that the preference of the principals is to do design work, rather than to seek commissions or administrate projects or the whole office. Questions arise as to whether the creation of good design can be managed on a larger scale by non-designers, and whether design credibility can be offered to the right type of clients if the practice is so large that attracting clients leaves no time for the best known senior principals to actually design.

Ahrends Burton and Koralek

Description in 1991

Ahrends Burton and Koralek, known as ABK, are well known and respected in architectural circles world-wide. They are a London-based practice, founded in 1961, and employing around 50 people. Their numerous and varied buildings have been and continue to be described in the technical press, and have achieved many awards. They were chosen as a case study example as they represent a well established, fairly large private practice with a strong focus on building design but no exclusive building type specialism – a healthy productive generalist practice. This is the 'norm' that was assumed to be a likely target for many private architects at the time of the RIBA study in 1962, and still is a dominant model.

Origins

The origins of the firm also bear a somewhat textbook resemblance to the image that many people, within and outside the profession, have of how a practice should start: a competition win enabled friends to found a partnership; a flourishing practice grew up and has continued for 30 years. The three founding and owning partners, Peter Ahrends, Richard Burton and Paul Koralek, met in 1951 in their first year as students at the Architectural Association in London. They had decided, before graduating, that they would try to form a partnership, the normal format for private practice at that time, as soon as they could obtain suitable commissions. Paul Koralek, while working in New York with Marcel Breuer, wanted to play his part in helping towards their goal of a joint partnership, and entered the open competition for the Library at Trinity College Dublin, which he won. In 1961 the project became a real commission, the basis for a new partnership, three architects and a secretary, working in space sublet from a quantity surveyor.

Development

Location

ABK have been in continuous operation since that time. In the first years, growth led them to move several times, first when serious work on the Trinity Library started, to Carter Lane for a couple of years. In 1970 they moved, with a staff of about fifteen in all, to their present North London premises in Utopia Village, Chalcot Road, conveniently close to their three homes. This space has accommodated them when they have been as large as 50 people, and they have no plans to leave it.

Practice activities

The service that ABK provide is primarily the design of buildings and co-ordination of their production, the 'classic' architects' output. However the firm has an interest in a wide range of activities. It has taken a special interest in energy conservation.

Partners have played an active role in the educational field, as when Peter Ahrends worked on a part-time basis in the office while the head of the Bartlett School of Architecture at University College London, and recently while Richard Burton has chaired the RIBA Education Committee during its preparation of a report on education. Richard Burton is also a director of a small company called Building Use Studies, with a social psychology special-ism, relating to buildings as they are perceived and used by the occupants, although the work done by this company is entirely independent of ABK.

Other activities include Peter Ahrend's membership of the Design Council, and Paul Koralek's advisory role with the Cardiff Bay Development Corporation.

Size and turnover

In 1991 the office had 50 staff, 40 of whom were either qualified architects or at some stage of pre-qualification, and thus fell into the band of the second largest type of office: those with between 31 and 50 full-time architectural staff. The present proportion of support staff to architectural staff at ABK, about 1:4, is somewhat higher than has been the case at other times in the practice's history. The staff numbers have increased and decreased over time, as projects have changed. A large project in an office of this size makes a very noticeable difference to the staff numbers, and runs of bad luck have similarly affected numbers. The policy for the partners to maintain a design role in projects, and for each project to be under the direct control of one of them, has acted as a natural limiting factor to the growth of the firm.

Projects and clients

The pattern of projects undertaken by ABK began with mainly public sector work, largely educational, housing or 'public' buildings such as libraries. In the 1970s the principals perceived an increase in work for the private sector, though the decade remained largely one of housing projects such as those for Basildon Development Corporation, and work for educational establishments. The cancellation of a housing project in Milton Keynes in 1980 seemed symbolic, as public spending was curtailed when Margaret Thatcher became Prime Minister. At that point commercial projects became a more reliable source of income; several large ones were started in the early 1980s and others began throughout the decade. Today the type of projects in hand or recently completed is more evenly mixed.

Projects have come from a variety of sources, which is consistent with the fact that the practice is not seen as a specialist in one particular field. Some early commissions were passed on to the practice from more established firms, a tradition that Richard Burton sees as typically British, or at least very un-American, and that ABK carries on. There are a great many projects which are either further work for the same client, or have resulted from a contact through a previous project in the same location. Out of 84 projects listed in a recent summary, eleven were competition entries, three of them winning ones, and of the remaining 73 almost half were for clients who returned for additional phases or other projects after the first project. Generally, repeat work has come from the public or educational clients, by their nature purchasers of more building work over time than most commercial organizations. The scope for repeat work in the private sector is greater for property developer clients. The practice has not yet done a large amount of work for developers, and until recently these organizations were not associated with the type of concern for design that is at the heart of all ABK's work.

Practice behaviour

Company structure

Choices about the ownership, hierarchies and legal existence of any organization have to be made many times as it develops. The three original partners remain the sole owners of ABK, but other things have changed. In 1978 an unlimited liability service company was formed, which employed all the staff and entered into all financial transactions with suppliers. This was converted into a limited liability company in 1985 after the change of rules by the RIBA made it possible for architects to operate as limited companies. The partnership was retained for the sake of those clients who might prefer to employ a partnership, often because they were clients in very early days, but the majority of the work is now done through Ahrends Burton and Koralek Ltd.

When the practice started, with its young partners, it had to employ people of more or less the same age and experience as the partners. Eventually, in 1974, three of these who had been there a long time – Paul Drake since 1962, Patrick Stubbings since 1966, and John Hermsen since 1971 – became associates and then salaried partners. Since then the firm has had to consider the structure again. About six years ago it addressed the consequences for a firm with six partners all in their early 50s and the need to promote some younger staff so that eventually they would be in a position to take on some of the partners' roles. So in 1986–7 three younger people, David Cruse, Christine Price and Jeremy Peacock, who had been with the firm for some time, became associates.

Staff selection and career patterns

Little advertising is needed for staff in times of expansion, as there is a continuous stream of letters from people seeking employment in the office. These are kept and responded to at such times as more staff are needed. There is a high turnover of architectural staff as there is, in the present structure, not much of a career path for the more dynamic individuals, who leave to head their own practices or to become partners in other firms. There is an occasional lunchtime educational lecture for the staff which is well attended, but little conscious career structuring. The work experience, in design and on site, forms the base from which staff move on. As CAD becomes more integrated into the office – it has been in use for about two years – there will also be a need for CAD ability and experience, and with increasing management tasks the office will also look more frequently for management skills and experience.

Management issues

Practice management takes the form of regular monthly meetings of partners and associates, and smaller informal meetings in between. Chairmanship of the monthly meeting rotates to one of the six partners for about a year at a time.

Procedural management problems for management of paper are becoming more difficult. The procedure for checking drawings is becoming increasingly important, both because more clients demand that drawings be signed off, and because of the eventual requirements of quality assurance. The task of checking drawings is skilled and painstaking, but can no longer be done directly by one of the partners as it was originally.

More formalized processes and procedures are being gradually established. In the last few years standardized specification clauses have been prepared and entered on the word processing system, available for use or for modification to become suitable for a particular project. Expertise on what should be prepared for entering a competition has been drawn together. Procedures to make technical information assembled for particular projects available to other projects in the office have been devised.

Financial management

Since about 1966, as its affairs became too complex to be run by a couple of secretaries, the practice has employed someone with financial training to help manage that area. Day-to-day needs can dominate and push routine monitoring and programming into the background, so more systematic procedures have been introduced, and ABK now have an accountant director of the company as a financial co-ordinator.

When considering profitability of projects the target for management of costs within the fee budget has changed. A larger percentage of the costs are now assumed to be committed to the overheads than to salaries, partly because there was a rise in rent, and partly because of the rise in indemnity insurance.

Marketing

The attitude to marketing is only gradually changing. In the past only one brochure summary describing the work of the practice has been prepared, in 1982. A monograph about the practice has recently been published, which includes a description and pictures of a large number of the projects that they have completed. An important form of marketing is through the publication in the professional magazines of illustrated articles about the buildings as they are completed. Another more indirect marketing effort lies in the spin-off from the competitive proposals that are an increasing part of the way in which clients seek a suitable architect. A more active approach to marketing has started, since adverse publicity over the National Gallery project led to a severe falling-off of enquiries. At this point they employed professional advice which has helped to turn around the negative press.

The first attempt at commercial advertising has just taken place, an experiment with an advertisement in the medical press at the time of the completion of the Low Energy Hospital which can be expected to be of great interest to this specialist field. It is a natural avenue to explore if attempting to capitalise on the enormous level of specialist skill that this project has had to build up.

Other associations and roles

The practice offers architectural design, interiors, and urban design and planning. This means that it does not employ, for example, quantity surveyors or engineers. For most of the other specialist skills that are offered by 'multi-disciplinary' offices, consultants are brought in. Partners discuss the current performance of advisers on a regular basis, so they can all keep up to date on which are likely to be suitable for future projects.

In addition a deliberate move was made in the past to have an in-house non-architectural specialist in the field of understanding user needs in housing. A social psychologist was employed by the office for three years while the staged design and building of the Chalvedon housing at Basildon was taking place. A similar process was followed to understand user needs for the Cummins factory. These initiatives confirmed the practice's commitment to the value of feedback in design, and became a trigger to joining with DEGW to set up a separate research company, Building Use Studies (BUS), to investigate and advise on the interaction of buildings with the behaviour and needs of users (see also Chapter 8).

The partners have always maintained many contacts with the world of architectural opinion-making. They have all had roles in schools of architecture as teachers and external examiners. They have lectured and published articles in the RIBA journal and other influential outlets. In addition they have played a role in the determination of quality in design through acting as competition assessors, and panellists for awards. Paul Koralek is a member of the Royal Academy.

Philosophy

'The ethos of the office is that the energy goes into design, not into management.' Statements in the summary of projects listed in the 1982 brochure indicate the firmly held and unchanging belief of the practice that the partners are there to design and to see their designs brought to fruition, and that this is what their professional skill and responsibility are about: 'Our organisation is based on project teams, in all cases headed by a partner... We insist on a high degree of partner involvement in all our work from inception to completion.' They want to explore different structural types and building types and seek to have the skills to do this.

Benchmarks for the practice are seen as commissions for new building types, or explorations of new structural and design possibilities. Perhaps the original partners were not only skilled but also fortunate to have started their own firm in the 1960s when life was optimistic for architects, providing opportunities to prove the assumption of their education: that a well trained and talented architect can design anything.

Notes

1. 1988 these two bands together covered 69.2 per cent of the total number of practices and accounted for 21.6 per cent of the staff employed in full-time architectural practices.

2. *RIBA Interiors*, September 1986, pp. 22–25.

3. To another design-based firm, Contract Interiors Ltd, whose director had been a contact of Stephenson's when he worked at BDP.

4. Manser, J. 'Practice profile of Stephenson/Mills', *The Architect, Journal of the RIBA*, September 1987, pp. 28–31.

8

New directions

DEGW Group

Description in 1991

DEGW Group Ltd is now internationally known as specialists in the design of the workplace. They have five established offices, in London, Glasgow, Milan, Madrid and Paris, with recently formed ones in Amersfoort, Athens, Berlin, Brussels, and Munich as well as Manchester. The headquarters of the group is in London. They have always concentrated on a type of design work that is somewhat out of the mainstream of building design, and as they have expanded they have extended the range of client services or products which have developed from their initial interests. They are multi-disciplinary, but not in the well known architectural sense of providing a complete in-house integrated building design service incorporating the range of engineering and costing skills. Instead their range of services focuses on the needs of end-users for interior planning/management assistance. In the fields of space planning,[1] building appraisal[2] and facilities planning and management they have been pioneers in the UK.

Origins

DEGW had its origins in 1971, as the London-based offshoot of a New York firm of space planners, JFN Associates Inc. Eventually after various organizational evolutions the partnership has become DEGW Group Ltd. The founding partners Frank Duffy and John Worthington, and a later partner Peter Eley who joined them in 1976, studied together at the Architectural Association school in London, graduating in 1964. While they were studying, many projects were carried out jointly by these three, and after graduating, as well as gaining experience in a range of different offices, they all spent time in universities in the USA on Harkness Fellowships. During this period Frank Duffy developed his interest in organizational theory and the design of office buildings and forged his links with JFN which provided the jumping-off point for the practice. The group of five original partners, all connected either through their education or through the space planning work of JFN, formed the basis of the present company.

Development

Location
The first office in which the infant practice took a share of space was in one of the large rooms above the Wigmore Hall. Growth led to several local moves. In the later 1980s explosive growth led to leasing a collection of small spaces in the immediate vicinity; this was a split location, with different groups being shuffled from place to place in attempts to accommodate changing project mixes and organizational strategies. New premises, big enough for the entire London operation, were urgently sought. In Porters North, at King's Cross, DEGW's location since February 1989, the opportunity was taken to use the new office to demonstrate the workplace strategy and the design expertise of the practice.

Practice activities
The reputation of DEGW has not been founded on the design of the buildings that they have produced. Initially the specialist North American skill, space planning, was of great importance. In 1983 DEGW described themselves as architects and space planners. A later brochure listed a formidable range of skills. The order of this list in part reflects an attitude to the type of practice activities that they expected: before definite building proposals can be considered there is a discrete research activity, information gathering and organization into briefs, that can exist independently of a design commission. Then follows architecture: a building is designed, either overall or in its interior, – in general layouts or in details of the products, and signs and graphics are used. Lastly comes evaluation of the success of the design, in relation to the users of the building rather than in relation to the architectural profession's assessment of the style and integrity of the design concepts.

The partnership deed for the restructured company in the mid 1980s stated that a prime responsibility was to 'ensure that the practice is based on a continuous quest for new ideas and procedures that are in the consumers' interest'; the implication was that the ideas would be not about innovative structural or design solutions, but about better understanding of how to relate user needs to design decisions. The distinctive character of the firm, whose reputation is built on the prolific innovation in concepts generated by the interests and expertise of the founding partner Frank Duffy, developed through heavy emphasis on research into organizations and their relationship to the space they use.

Size and turnover

The company grew more or less gradually in the early days, increasing to about 25 staff by the time they moved to Welbeck Street in 1975, and to about 50 by the time they moved again, to Bulstrode Place. At the time of each move a short period with little or no growth followed. The really dramatic growth in numbers took place between 1985 and 1990. By 1989, when DEGW moved to King's Cross, there were over 150 staff in the London office alone. This was largely the result of great expansion of office building and replanning activities associated with the boom of the late 1980s, and the anticipated changes in the finance and 'Big Bang' related markets. DEGW developed and refined the services it could offer to office building developers, owners and users just at the moment when people were crying out for guidance in this field. At the same time the now maturing office received commissions for large interior design projects, and undertook some new office building projects. Foreign contacts were broadened, including focusing attention on Japan. As well as this the office was continually adding supplementary activities to the main strands. Many small additions added up to a significant amount of growth.

In its forward-looking internal conference, Futures, in October 1988, set up to explain and explore the prospect and intentions of their proposed incorporation, continued rapid growth was forecast, looking towards a group over twice as large by 1995. What actually happened was the opposite. A number of circumstances coming together meant that, earlier than most, the firm suffered the financial problems that have subsequently hit all companies during the current (1991) recession. The taxing experience of incorporation, an expensive office move, a severe overrun on some projects and the backlog of smaller unprofitable projects, coupled with a reduction in the profitable fit-out and interior design projects, led to the need to cut back staff dramatically.

Projects and clients

Publications have brought DEGW into the public eye, starting with *Planning Office Space*, published as articles in the *AJ* in 1973, and as a book by the AP in 1976.[3] The next major opinion-forming document was *Orbit 1* in 1983–4, followed by *Orbit 2* in the USA in 1985.[4] The former set out to investigate the impact on office design of information technology (IT), the genie that was about to burst out of the bottle. The conclusions about the premature obsolescence of buildings which could not adapt to accommodate IT were enormously influential, especially in the developer world, and led to their building appraisal methods, the rapid increase in the amount of work done for the developer clients, and the start of an ongoing set of client sponsored research projects. One of the key characteristics of the firm's workload is that a few large projects are usually accompanied by an enormous number of very short-term ones. Many of the projects that are carried out are short and sharp, perhaps taking two to six weeks. There is a management challenge in co-ordination of the rhythm of the short ones – often the work of one or two people, whose output is usually an A4 spiral-bound report – with the long ones, which may take many months, and will have to handle large CAD drawings, incorporate regular design team meetings with large casts, and be subject to the typical unpredictability of building projects. The number of reports prepared in a year can be taken as an indication of the number of projects completed. Since 1978, there have always been 60 reports each year. In 1978 the number was 78 and in 1985, one of the busiest years, it rose to 119, but in 1988 it had dropped to 67.

Practice behaviour

Company structure
The structure of DEGW has undergone a number of changes, in particular in the period immediately before and after incorporation. This has been related to managing the overall growth, the diversity, and the many different types of service that are offered. In time there were nine partners, all from design professions, with responsibility for different groups of staff and generally somewhat different project types depending on their skills and backgrounds; and there was a managing partner, one of the senior partners, to look after the affairs of the company as a whole. This is a familiar structure. With minor modifications it lasted until the late 1980s. With growth the partners became concerned to create structures that could potentially outlast them as individuals, and looked at the possibility of incorporation to help them achieve this. Thus the emphasis shifted to a structure focusing on the different types of work, and the different groups within DEGW, rather than the individuals. The structure continued to evolve, struggling to find the right balance between functions, people and office identities, and has now settled on the concept of a group, DEGW Group Ltd, which sits above the local offices, leaving the London office, still much the largest, as a local office with its own managing director like all the others. The London office is divided by product, which now excludes for example research but includes building appraisal and strategic facilities planning. The company DEGW T3 deals with product and graphic design. The unit heads of each group, as well as the managing directors, are responsible for their own profitability and cross-charging of staff when they work within each other's units.

Staff selection and career patterns

A new partnership deed written in the mid 1980s placed the responsibility 'to have a concern for the professional and personal development of every member of the firm' first, followed by that to 'provide a service designed to meet the needs of different client types'. Staff with a wide range of skills were needed. During the 1980s a part-time academic appointment within the firm was made to ensure that continuing professional development was well provided for.

Staffing ratios have changed over the years. The ratio of support staff to design professionals being progressively tightened. In the early days of the practice the ratio was 1:1.3 support:professional; that improved and it is currently 1:4 in the group as a whole and 1:4.5 in the London office.

As the company has grown the role of staff development has shifted to the units, and varies in how it is managed within each unit's budget. The role of in-house lectures has also shifted. The conference facilities in the new premises are of great significance to the organization, though the events that take place there are often of a semi-public nature with well known outside speakers and outside guests; there are even public announcements of events open for anyone to attend. These changes embody an imperceptible shift which reverses the priority of the statements in that partnership deed.

Management issues

Management has always interested senior members of DEGW, whose first point of departure in design projects is to consider the organization for whom they are designing, and its management style. How to manage their own organization, however, still proved a difficult nut to crack. Although not reluctant to consider the need for professional management, the partners did not seek training in management skills themselves, but eventually selected a chief executive with a track record in management, though with an architectural background as well. Project management of the larger projects by individuals employed only for that and with specialist skills in the field was established earlier, but did not influence the general approach to management very significantly.

Day-to-day management within units is now handled by each unit leader, who has to plan resources and the progress of projects, and is fully responsible for the profit within the unit. The central group unit manages joint affairs and some aspects of publicity, but there are few centralized services to assist day-to-day operations, which is a change from the period immediately before incorporation. The very widespread use of fully networked computers means that each unit handles its own paper output on standard formats and develops its own approach to quality control of that output.

Financial management

Financial management was if anything even more difficult to put on to the right track than general and project management. The outside firm of accountants on whom much reliance had been placed had to be matched by internal expertise as tax and financial affairs became more complex. There is now only one non-executive director. His presence at board meetings, and his contribution to financial understanding, has continued to improve the focus on the business side.

Marketing

The practice has a long history of publicizing their ideas, particularly on office design. Since the publication of the articles about office design in the *AJ* in 1973, articles, books and lectures have been a major part of the activity of especially the senior partners/directors. The desire to emulate a university environment, to share insights and generate intellectual debate, has been a strong motivation, but the value of these activities in keeping DEGW constantly in the public eye has been equally important. The balance that needs to be struck between sharing information among equals, and giving away a business advantage by letting rivals understand your methods, has always been a delicate one.

Alongside these aspects of marketing has been the development of a formal marketing approach: systematically getting publicity for projects in magazines, preparing increasingly carefully targeted brochures for handing to prospective and existing clients, and inviting influential people to visit the office and to attend lectures. The central group has an important marketing role, and new approaches are regularly tried: currently well over 2000 'clients and friends' receive a monthly fax of DEGW news

Other associations and roles

DEGW has an unusual list of activities and other disciplines, companies and enterprises with which it is concerned as well as the various internal disciplines. Small companies have been embodiments of activities close to architecture, space planning and interior design, whose development has been sponsored by DEGW in input of ideas and money. An example is the share in the creation of a company to investigate, on a social psychological basis, the behaviour of people in buildings, namely Building Use Studies (see also the previous chapter on ABK). For the first years of its existence this company shared offices with DEGW, moving to separate premises before the move to King's Cross, and gradually becoming effectively separated, though some ties remain. Another activity that grew into a small publishing company, Bulstrode Press, was the editing and production of a monthly newsletter for facilities managers. DEGW T3, the product design company, has already been mentioned, and currently forms part of the group; and a new association with an information technology consultancy company, Effective Technology Ltd, led to DEGW ETL recently being formed.

There have been many members of staff who have had part-time teaching roles generally at architectural schools in and near London and even in the University of Cincinnati in Ohio. Other activities have involved Frank Duffy playing an increasing major role in the RIBA, of which he is currently (1993) president, and making many television appearances. Planners in the urban design unit have taken on the organization of the Urban Design Group, involving a wide variety of joint research projects with organizations such as Butler Cox, Eosys, and the Building Research Establishment (BRE). There has been in all ways a constant outward-looking approach, a seeking of ways to widen the network of interested contacts.

Philosophy

The practice has always believed that the needs of the user are the starting point for design, and that design should often concentrate on the inside as well as the outside of a building. It has also increasingly come to consider the importance of the management of buildings after they have been completed and handed over to the client, as distinct from the fabric and building systems themselves. For this reason DEGW has not been troubled that its reputation does not lie in the area of designing new buildings. Rather it takes pride in also being concerned with the renovation and refitting of existing office stock, as something that is in constant demand. There has also been an ethic that has assumed that knowledge is to be shared. As the business climate has hardened this has been harder to keep distinct from giving away secrets.

Building Design Partnership

Description in 1991

Building Design Partnership (BDP) is the largest practice in the UK. The largest single employee group is architects, 440 in 1989, but there are also landscape architects, planners, interior and graphic designers, quantity survey- ors, civil, structural, mechanical, electrical and health engineers, as well as project management specialists. In the late 1980s a major output of the firm was retail and office building, with planning and industrial work next in their share of turnover, and with much smaller contributions in health education and housing. This is being replaced in the 1990s by public sector work and major transport projects.

Origins

The firm developed from the practice started in Preston in 1937 by George Grenfell Baines, known as 'GG'. Grenfell Baines's early experience left him with a great respect for the skills and knowledge of the different professions associated with architecture, the various engineers and surveyors. By 1941, the work generated by the war brought together three separate one-man companies, and led to the growth of a group of eight in a multi-disciplinary practice of architects, engineers and surveyors, the Grenfell Baines Group, where income was decided by mutual assessment of the contribution of each member of the group. The development of the present firm was gradual but continuous, through a number of different entities and combinations, until in 1961 BDP was formed, the first real multi-disciplinary design practice in the UK. This legacy from Grenfell Baines, and the profit-sharing approach from the early days, have remained central to the way in which BDP works.

Development

The history and growth of the firm spans from 1937 to the present day, covering the period of the Second World War, the birth of the National Health Service, the arrival and demise of Parker Morris standards in housing, the explosion of technology in building methods as well as in communications and data processing, and many other events of note. Unlike the other cases, BDP was well established at the start of the period covered by this study, and some mention must be made of the developments before 1968, which set a context for the most recent twenty years of growth and adaptation. In 1987 BDP celebrated both its Golden Jubilee, 50 years for the practice, and its Silver Jubilee, 25 years as BDP. For that occasion a detailed documentation of the history and development of the firm was prepared by retired partner Bill White and published as *The Spirit of BDP*, [5] from which much of the following is distilled.

Location

The first office was in Preston, Grenfell Baines's home town. When significant projects have been located elsewhere, or when key people's residential preferences have taken them away from Preston, other offices have been opened, and in some cases closed again, in numerous locations, many of them in the northern half of Britain. None of the earlier offshoot offices are now important, and their closure receives no mention in the 1987 historical review. In 1959 the 22-year-old northern firm first opened an office in London, in a flat in the West End. By 1961, when BDP was formally born, they were established in Cavendish Place. In 1968, and thus for the period covered by this study, an office was opened in Gresse Street, where they now occupy more than one building. There have been flirtations with offices abroad, though they all had relatively short lives, often having developed around particular associations and projects and being wound up when those came to an end.

The London office was different in kind from the rest of the firm. It was headed by David Rock who had come from Basil Spence's office, and was appointed by GG to start an office in London. He was joined by Bill Jack and Bob Smart, both now retired from senior positions in the firm. In its early days the London office focused on design issues, and submitted entries for and won competitions, yet did not figure prominently in the overall picture that BDP had of itself. Each office culture reflects its location and the influence of the staff based there. Central activities are located where the relevant individuals are based. In 1991 the Chairman was located in London and the Chief Executive in Preston.

Practice activities

Design, rather than architecture, is the service that BDP wished to offer from the very beginning. Within the group of technical fee-earning staff, this means that a wide range of design professions are represented, the four major ones being architecture, mechanical and electrical (M&E) engineering, civil and structural engineering, and quantity surveying. In 1971 this list included, in addition, presentation staff and administration. By 1990 the list was much longer, including landscape architects, interior designers, graphic designers, planners, as well as project planning and management specialists, and CAD and computer support staff, all with direct involvement in projects. These various disciplines are broken down into even more detailed specialisms in some instances. The inclusion of the various disciplines means that BDP can offer an integrated design service to clients, or alternatively can, and does, offer the separate services individually. At present the M&E services, a large component in cost, and influential in the design strategy for many of their projects, is relatively easier to sell as an additional or even as a separate service than the others. The disciplines with smaller numbers represented can be ranked according to the ease with which they are successfully offered to clients. Some very small groups are strong in niche markets.

The multi-disciplinary aspect of the practice is not a static characteristic, despite its long history. It is influenced and affected by shifts of attitude within and outside the practice. Clients for large sophisticated projects today tend to employ QSs directly, to act on clients' behalf as 'cost policemen'. This means that the services of BDP's own QSs are less frequently in demand as part of an integrated team, with the effect that their numbers declined at the start of the recession. At the same time there is active discussion about the best way to present BDP to the world. For some this involves presentation of different design skills separately, to indicate that BDP contains centres of excellence in these different fields, and to attract recognition as first-class designers.

Size and turnover

During the 1950s the turnover was relatively even, and staff numbers stayed around 50 for much of that time. As BDP was born there was a great spurt and development. It was nearly ten times as big in 1971 as in 1959, a huge expansion, a great challenge. During that period, the rate of growth in turnover was not quite as dramatic.

In 1967–72 the staff numbers were growing faster than the turnover, so that by the time of the recession of the early 1970s BDP had to shed nearly 140 of its 722 staff, about 19 per cent. Throughout the rest of the 1970s staff numbers grew only very slowly, and growth in turnover increased comparably. In 1982–3 100 staff out of 834 were lost, but by the mid 1980s expansion was well under way again. The late 1980s saw a huge increase in turnover, and staff numbers have also steadily increased, to 1485 in 1990. The current recession has again affected numbers. Staff have been lost from all levels, including partners, and while the effect is not evenly spread across all staff groups or offices, it is being felt everywhere.

Neither growth nor decline has affected all the offices or all the specialisms to the same degree. A BDP publication illustrating the changes between 1961 and 1981 shows the type of differences to be found.[6] The changes in numbers and balance of staff clearly reflect local patterns of demand and the specific nature of each local office and its way of responding to that demand. In some respects BDP is actually six or seven separate offices with their own characteristics, bonded in a group with overall directions, philosophies and methods in common, but acting out their professional lives in different arenas; they are described by one senior partner as a 'convoy of ships going the same way'.

Projects and clients

BDP has had many clients for whom it has carried out repeated projects. It has worked on the whole range of building types: universities, hospitals, commercial and government offices, retail complexes. Fundamentally the enormous growth of BDP, when contrasted to other UK firms, is due to the fact that its philosophy and consequently the way in which projects are managed are better suited to larger projects. Not many teams could take on in their entirety some of the projects that BDP has been running, such as the Channel Tunnel terminal with a value of £250 million. The firm needs to be big to be able to handle such projects as a whole.

Practice behaviour

Company structure

The offices are run as separate autonomous organizations, with their own control over projects. In addition there is a strand of management related to the different design professions, each of which has an identity within the firm. These two intersecting patterns have to be unified under the general umbrella of BDP. Now each office is run by its own group of partners, who set targets, market their services, supervise the design and management of projects, and hire and fire staff. The central organization contributes some central services to all offices. These have varied over time, sometimes contributing a scarce resource to be used as needed in any location, and at other times, as now, concentrating on offering administrative and strategic guidance. Central financial administration, involving common methods of assessing achievement of targets, is crucial to the unified identity of the practice. It is not through design style that BDP can be typecast, but through its method of delivering a service, and by its quality. The Chairman, these days charged with presenting BDP to the world, shaping its image and creating the necessary structures to reinforce that, is a single individual, who may originate in any design discipline but who represents all disciplines and all offices. The Chief Executive is a single individual in control of major management decisions for the whole group of offices. These two are elected by all the partners for a two-year term, a democratic process.

The issues of quality of life, staff career patterns, and the pattern of payment levels, are centrally determined. Profit-sharing, from the days of the early GB Group agreements, has developed into a way of setting salaries on the basis of shares in BDP. Staff are allocated shares (though this does not mean that they own a 'share' of BDP) according to their experience and performance. The pay that a person receives is a product of the share value and his or her number of shares, thus avoiding differentials between different offices.

Staff selection and career patterns

There is a career structure for all the professions, but the development of careers is within the separate offices. 'The control of destiny is profession by profession, office by office.' BDP recruits by profession, and is constantly concerned to raise design standards. A schematic five-stage career pattern can be seen in seven-year chunks, all of which could take place in BDP:

1. running oneself (newly qualified)
2. running jobs
3. running groups of jobs
4. running one of the offices
5. running out.

This would all add up to 35 years, from say age 27 when one might join after passing the RIBA Part 3 exam, to 62 and retirement. BDP has held internal conferences since 1960 on a more or less annual basis for partners, less frequently for associates, as well as on two occasions for the entire staff.

Management issues

Participation in the 1962 RIBA study on *The Architect and His Office* was an important stage in the development of the practice's attitude to management and also influenced the outcome of that study considerably. A number of the dynamic younger staff of that period, now in leading positions in BDP or their own successful practices, had early job experience in the USA and thus had a perspective of larger practices. These influences were not consciously drawn on to help structure the practice as it grew, but it seems likely that they were important. The firm has always stressed that excellent design should be accompanied by a well managed service, and so management of the firm and of projects has always been given a great deal of attention. However, some of those who have left BDP talk of themselves as people who put design first, as they didn't want to rise up a management/administration ladder to the top. Nevertheless, both the RIBA studies, in 1962 and 1992,[7] stress how vital proper management is. The pattern of management structures has changed over the years, and will continue to do so. The current structure is that of matrix management. Offices, jobs, professions, professionals, each have their own lines of control, with interlocking intersection points. There are a number of central functions which decide overall policies, and each office defines its own policy to implement central decisions. There are job structures which may call in resources within the office and within the firm as a whole. There are professional structures seeking to enhance each profession's profile and performance. The policies and proposals across each of these lines are implemented within each office. Quality assurance (QA) is being developed. A central unit is in charge of the process developments for QA, although it is the business of each local office to get accreditation. The four major professions – architects, civil engineers, M&E engineers and QSs – are all accredited under BS 5750.

Financial management

There has been active financial information available and assembled comparably across the different offices since BDP was officially formed in 1961. A senior financial manager was first included as one of the partners some twenty years ago. In common with the experience of other architectural companies who start to use financial management for the first time at a new level, initially without the skill and experience to evaluate and control it, this first experience was not a happy one. The difficulties that arose were the impetus for a set of financial controls and management information supplied to all partners, that has been of central importance since that date. The job costing system, done on an accountancy basis, has proved an important element in ensuring that even in a hard recession BDP has retained a good share of the available design work. The controls and systems enable long-term trends to be seen and allow a higher level of strategic discussion to take place. Managers can match expected patterns to reality and set about control and marketing in a well developed and conscious framework of information.

Marketing

In the early days marketing was not thought of as a particular function; Grenfell Baines was a natural marketeer. By the late 1970s a Strategic Policy Steering Group (SPSG) was given the remit to:

1. look at what BDP was doing;
2. look at what it could do but was not;
3. identify what it would like to do but was not yet professionally allowed to, e.g. act as developer.

There were seventeen different professional institutes represented by the partners at that time. A need for targeted marketing was identified.

In 1980 marketing experts from the USA, Gerre Jones and Weld Coxe, came and discussed marketing concepts with BDP. The firm became aware through this exercise that it had resources, in the form of people with specific knowledge and expertise in a particular field, who could be 'labelled' and who would instantly create their market, e.g. Keith Scott for retail buildings. This led BDP to realize that it had specialized knowledge of different sectors, which it must exploit. The identity of the expert resides in his ability to design for a particular market rather than his design style.

The role of SPSG became the function of the Central Marketing Group, and a co-ordinating role is currently played by Corporate Communications, while the direct marketing function has become the responsibility of local offices. These sorts of changes are caused in part by shifts between recession and boom and back to recession. The possibilities inherent in following longer-term management policies are disrupted by the discontinuity of the workload. The firm has looked abroad during recessions, e.g. to the Middle East in the 1970s, and is no doubt doing so again. Alliances already exist in France and Germany, new stronger links are being forged in Ireland, and Spanish and Portuguese contacts have grown in importance.

Other associations and roles

The merger between BDP and Lowe and Rodin, structural and civil engineers, in 1970 was an important step in strengthening these disciplines and providing several key individuals. Currently more specific specialists are being employed, as the firm is broad and big enough to warrant specialists with very particular knowledge, able to operate across the whole firm, for example: lighting, energy, materials conservation and environmental concerns.

GG has always had an interest in architectural training. A teaching practice was set up in Sheffield at the time when he was Professor of Architecture there. It was conceived on the model of a teaching hospital which is used to train medical students, in the hope that jobs in the office would provide a captive set of real problems on which students could learn. However as an idea it lacked the parallel to a medical clinic: there was not necessarily a body of clients as readily available. It lasted while GG was professor at Sheffield, and for a period under the aegis of BDP Sheffield, but it finally withered.

Philosophy

BDP do not believe in a cult of the individual; prima donnas come and go, whereas they wish to be able to continue through the generations presenting a well designed and efficiently delivered product. They are, after all, now embarking on third-generation partners, mostly taking over from the second generation, the founders having left some years ago. There is a perception by some architects, and propagated both by the media and by education in some instances, that the most creative designers are individualists who will want to run their own practices. This contrasts with the belief within BDP that 'the blending of talents is necessary if good, balanced service in this complex field is to be given... We have freed ourselves from the inevitable decline which faces a personal practice built around one or a few key people.'[8]

BDP sees its 'consistent long term purpose as the creation of greater satisfaction for clients and members of the construction industry by overcoming barriers and fostering team work and excellence of achievement.'[9] The stress is on the provision of excellence in both design and service, and there may be many different aspects to achieving the excellence for which they strive. The multi-disciplinary approach is a distinctive feature, a strength and a source of uncertainty. If a service is already complicated, as even a single-discipline architectural service must be, then, in times of rapid and profound change, how to guide this change for a more complex organization, how to redefine the nature and relationships of the different disciplines to achieve the high standards they set themselves, become truly awesome tasks.

Hampshire County Architects Department

Description in 1991

The Hampshire County Council Architects Department is a public authority architectural department that is one of the most highly regarded by the architectural profession for the quality of the designs it has produced, especially in the last decade in school buildings. It is headed by the County Architect, Colin Stansfield Smith, recently awarded the RIBA Gold Medal.

Origins

The origin of a department of this kind is not as readily identified as that of an independent office, as its existence is linked to that of the public authority, and it has a comparably long history. However, there was a major restructuring in 1972 of local authorities, which left Hampshire as an exceptionally large shire county, one of the three largest in England. Hampshire Council acquired a full-time leader, Freddy Emery-Wallis, and shortly afterwards in 1974 Colin Stansfield Smith became the County Architect. This combination of change was highly significant, almost enough to constitute a new beginning for the purpose of this case study, though the base from which changes were made by Stansfield Smith is in itself important.

Colin Stansfield Smith had been Deputy County Architect in Cheshire. From that background, Stansfield Smith was aware of the profound change that a determined chief can bring about. When appointed, he made it clear that he saw his role as bringing design, 'real architecture', into the department. He soon introduced a team of his own, key staff who followed him from Cheshire: John Robinson who became his deputy until 1979 when he left to become Chief Architect in Cumbria; David White who remained a key member of the department until his premature death in 1985; Huw Thomas as a project architect; and Mike Leonnard. These changes constituted an upheaval that was to have far-reaching effects on the development of the department, as project development and design methods were changed, and was to bring it to prominence as one of the foremost design offices in the country.

Development

Location
The County Architects Department is based in Three Minsters House in Winchester, and has been for the whole of the relevant period. The way in which a location might be chosen, the typical influences that might be found in

a private firm, the changes that occur as a firm develops, are not generally present for a public authority office, since its movements are interwoven with those of the whole council.

Practice activities

A local authority architecture department is responsible for the capital programme for new buildings and for maintenance of council owned and run buildings, in Hampshire's case, at county level. Unlike private practices, such departments have clearly defined limits to their authority: some activities are part of the work of the department, and others are not. If the Chief Architect wishes to go beyond these boundaries this must have politicians' approval. Changes in what the department is called upon to do will be related not only to the skill base it has built up, its interests or its previous development, as might be the case in private firms, but also to political views of the appropriate field of activity for the overall authority, with consequences for all providers of services.

However, for the period under review, the role of the department was both to act as provider of buildings identified as necessary by the services departments of the council, such as education and the police, and to take on the landlord's responsibility for the maintenance and upkeep of the stock of all ages. Thus the department combined staff to deal with the two parallel roles. The design staff included all the necessary specialisms such as engineers and quantity surveyors, for both the new work and the ongoing maintenance.

Size and turnover

The staff numbers in the department in mid 1990 were approximately 200 overall, including a full range of professional staff, most of whom are involved in the maintenance and running of council buildings rather than in new building projects (see later). Some staff work at home and others are based in area offices. The capital programme for 1989–90 for new-start projects was about £20 million. A separate programme of repair and maintenance is run by the surveyors, working under the County Architect, and for them the 1989–90 total expenditure was approximately £25 million. The size of the workload and thus of the department is linked to the expenditure programme for the authority as a whole, so the development of the department must be seen partly in terms of the required turnover.

In the 1960s there was a very large building programme, reflecting the enthusiastic rebuilding of Britain as recovery started after the Second World War. School roles were increasing rapidly in the 1960s, creating an urgent need for new school buildings, and the very fast turnover of projects at the time when system building was being developed led to standard plans and the use of the Second Consortium of Local Authorities (SCOLA) system. Changes in the requirements that the council must fulfil, changes in the balance between public and private investment, are the forces that affect the size of the programme, which today has declined considerably.

Projects and clients

The immediate client, the paymaster, for the Architects Department is the county. The real client is the user public, for whom it provides, for example, schools, courthouses, police and fire stations and other publicly funded buildings, but not housing.

There are other parts of the council, and the thirteen district and city councils, all of which have roles to play in the provision of publicly funded services and premises.

Practice behaviour

Company structure

The most significant change introduced by Stansfield Smith into the way in which the department worked took the form of a structural reorganization. There had been teams, of about twelve people, that concentrated on particular building types, acquiring an expertise in say schools or fire stations, which was thought by his predecessors to be a necessity when the department was under the post-war pressure to supply an enormous stream of buildings very rapidly. The teams were supported by a small group including a librarian and a typing pool. Stansfield Smith believed that 'a specialism can soon become stereotyped, prescriptive and prejudiced'; design may become based 'on assumption and formula', and thus can suffer from the lack of the 'innocent questions'. So he split his department into three geographical divisions, central, north and west, each with its own team of administrators, each with its own chief and deputy. Among the divisional heads were people that joined the department as appointees of the new Chief Architect, an effective way of producing a rapid change in the outlook and aspirations of the department. One other group, under David White, was created as a resource division to take on selected projects – 'jewels' – and set the tone for future projects. This group was dissolved in 1984 when its relevance had declined.

The area-based structure was intended to enable people to become familiar with their area, its needs, its existing building stock and its local materials, and to respond to its culture on a wider basis when designing a building of any type. Staff opposed to this way of working left the department, some immediately, some gradually. It has been restructured since into four operational divisions, but the important restructuring, the early grouping by area, had a long period of successful working.

Staff selection and career patterns

There has been a big shift away from using technicians over this period. Over 40 architectural technicians were employed in 1972, falling to six or seven by the late 1970s. There are now only two senior technicians, who have been with the department since before Stansfield Smith arrived, and whose whole approach is somewhat different from most technicians, as they are interested in the challenges of the new attitude to design. Occasionally a few junior technicians are employed doing working drawings, but frequently small local practices have been used to enable a rapid working drawing phase to be accomplished, though this practice may not continue.

Over this period there has effectively been an almost total change of staff. Good young designers have been anxious to join the department, and have stayed for a few years to gain experience before setting up on their own, and often locally, sometimes taking on some smaller projects that the department needed to give to outside contractors. Overall the average age of staff has been getting younger.

Management issues

Professional staff in the department are local government officers. Conditions of service of staff are part of the national conditions set by the National Joint Council for Local Authorities. A document is given to all new employees setting out all the conditions. They cover such issues as flexitime for certain levels of staff, sickness procedures, pensions, car use and so on. This is somewhat more formal than would be found in the average comparable private architectural office, and many of the conditions reflect a whole nation's way of working in public authorities, though there are some special provisions specific to Hampshire. Management of the work of the department, however, is the province of the Chief Architect and his management team.

Financial management

Since about 1986 the department has had a financial manager, Andrew Smith, who is really effective. Final accounts tend to be done by the in-house QSs, and until recently they carried out the accounting processes for the department's projects as a whole. However a greater management role was needed to ensure that the department would not become vulnerable from poor financial management, and also to help increase the range of potential opportunity by seeking out cash wherever it may be obtainable. Andrew Smith, an accountant appointed by the County Secretary's office was given the role of financial manager for the department, and he has taken a fully participative approach in the department, preparing financial reports for the council committee on individual schemes and on the overall department budget.

Marketing

In the early days of the period under review, marketing was a very different issue for a public authority department than for private architects. The problem was to provide as and when required, to a good standard and within budget, rather than to attract clients. This has dramatically changed, more radically for the public sector than for private practices. The decline in the public building programme has meant that there is simply less work available, barely enough to keep design teams busy in some instances. To maintain teams that have been built up, work may now need to be sought outside, and that involves marketing in the way that private firms do. The current description of the work of the department provided for new employees states that divisions may undertake commissions 'on an agency basis', and the Hampshire department enters competitions for outside projects.

Other associations and roles

The department employs architects, building and land surveyors, quantity surveyors, clerks of works, engineers (electrical, mechanical and structural) landscape architects, interior designers, graphic artists, historic buildings specialists, a model-maker, and administration staff. These cover the two strands of the department's work, one for management and regular maintenance of the estate, and one for creation of new buildings or major alterations or repair upgrades of existing ones. In the period considered, there was considerable use of consultants in the new building programme for specialist functions, and some judicious use of outside architects, in particular some well known nationally or even internationally for their design skills.

Philosophy

Colin Stansfield Smith has very strong beliefs about the role of design, the ways in which to get good design and the responsibilities of a public department of architecture. 'Buildings create part of the well-being of Hampshire.' He wants to enable people to identify with an area or region, and fears that proposed changes in local government could eventually mean that there would be real local losses of knowledge and tradition – not just materials, techniques or textures, but the chemistry of an area. He also takes an active role in trying to bring artists and craft skills into the design arena.

He is fully confident of the architect's ability to contribute uniquely to answer vital questions about what and how things should be built. However he is also fully critical of the levels of mediocrity that have been offered to the community, often by public authorities, which he sees as having an inescapable responsibility to do better, whatever the different pressures that may have to guide the more commercial producers of buildings.

Pentagon Design and Construction

Description in 1991

A comparatively young firm, formed in 1985, to carry out design and build projects. There are six directors and around twenty employees. They carry out projects in the medium-size bracket, with contract values of £2 million to £20 million, and unlike some other design and construction companies, they do not offer any other type of construction services. They were chosen as a case study to provide an example of another view of the building procurement process, looking not at a construction company offering a design and build service if required as part of the sale of its construction services, but at a company offering this form of procurement alone. None of the partners are registered architects. Their role in relation to the profession is as the employers of architectural firms, commissioning their design services on behalf of clients, within a framework agreed between themselves and the client. This is an increasingly important way in which design is commissioned, and is one that some architects find threatening.

Origins

The group was formed as a result of the joint working experience of the majority of the founding directors in the design–build part of Conder, the steel manufacturing company, Conder Southern. The founders have all had experience and training in construction-related disciplines, three in architecture as technicians, one in structural engineering, and one in quantity surveying. Two were involved in their own architectural and design firms at earlier stages in their careers. In addition they have between them had exposure to timber, concrete and steel specialist firms, often in the sales capacity, and thus have knowledge of the supply side of the construction industry.

Development

Location

The firm started in Arlesford. One of their projects, in Southampton, offered an opportunity to take on the building they currently occupy, the White House, in Grosvenor Square, which they refurbished as part of a large, mainly new-build office project. They moved there in June 1990. These premises are designed to create an image of a prestigious and solidly founded organization.

Practice activities

All the projects that they undertake are organized as design and build contracts. Their selling point is that they offer the client a building that meets the brief, the budget and the time frame. As part of this service they can offer a maximum guaranteed price (MGP), so that should there be an increase in cost they bear the loss, whereas should the subcontract figures eventually mean that there is a reduction in the price the client will benefit from this reduction.

Size and turnover

In order to manage their work in their peak year of 1990, when the turnover was £23 million, they had 22 staff. This had not changed much by 1992 despite the general down turn in construction as a result of the recession. There are, in addition to the directors, four quantity surveyors and four project managers, one accounts clerk, one construction manager, one architectural technician (to do the 'architectural perspectives' that they offer to their clients as part of the service) and three secretaries.

Projects and clients

They have mostly been involved with offices, for both developers and clients intending to occupy the buildings. They have also done some residential buildings, and are about to start construction on a hotel. In some instances their architectural contacts have recommended them to the client when a contractor was being considered. It is also their justifiably proud boast that in all cases their clients have come back to them for at least one more building. They prefer not to spend time in competing for business, although this is currently hard to avoid, and do not put in much effort whenever there are more than four firms being considered including themselves.

Practice behaviour

Company structure

The structure of the company, in funding and financial terms, has had a more complex history than many architectural practices or companies of this size, reflecting its different position in the market-place, and the different experience of its founders. Initially they sought financial credibility and respectability by involvement with an established organization, and A. Monk and Co. PLC took an equal share of the initial investment with the five founding partners. In 1989 Pentagon changed their status to call themselves a group, Pentagon Group Holdings when they joined forces with Tribune Properties, a company that had been created in 1987 to act as developers where suitable opportunities arose. A dormant organization, Pentagon Homes Ltd, was created to enable NHBC certification to be obtained, in readiness for the time when the housing market should become active. This comprehensive business approach is rarely found in small architectural practices or other companies.

In 1990 A. Monk and Co. were taken over by Davy Corporation. In 1991, when Davy were themselves to be bought out by Trafalgar House, Pentagon decided to take the opportunity to buy back their shares from Davy. This was negotiated successfully, so that now the directors own the major part of the company themselves.

Staff selection and career patterns

There is a strong emphasis on selecting staff with construction experience, both for management roles and for quantity surveying and cost work. Architecture and engineering are considered to be well represented by the directors so that there is no recruitment of in-house staff in these areas. The process is to hire outside companies for both these areas of design, to choose the best for the project and to supervise them closely, but to avoid the ongoing overheads of such staff when projects do not require them.

Management issues

Site management is of considerable importance, so site offices are established that are well equipped for distance work, with faxes, intercoms and mobile phones, and the office staff have car phones. They often employ freelance site managers, enabling them to keep their overheads down. The firm tries to establish with site managers that they should not, as is often the case when they are employed in construction firms, be seeking to find opportunities to blame the design team for things when they go wrong. Their approach should be one of problem-solving rather than establishing blame.

Financial management

Unlike many typical architectural firms, Pentagon had a financial manager on board before they started the company. Their financial director (a QS) joined them after the company had been started but had acted as a consultant in the early stages of setting up. This attitude – that financial management is essential – is fundamental to their management and working pattern. The maximum guaranteed price is an aspect of their 'product' that they believe to be a strong selling point, and to achieve this at a profitable point for themselves and their clients requires financial control and understanding to be central to their operations.

Philosophy

Design–build is a specialist service that Pentagon believe has a place in the building procurement market. They do not attempt to sell the service either to provide work for their own construction team, or to market a product that they manufacture – both common forms of design–build enterprise with which they have had prior experience. These approaches, they feel, are not in the true interests of the client. They are aware that designers are often wary of the design–build approach, on the basis that they will not be able to have a free hand in the design, and may not be able to use their talents to the best advantage of the project. The company do stick to a budget that has been created at a very early stage on the basis of informed opinion of what can be achieved in cost per unit area ($£/m^2$) for the client's benefit (as well as in their own interests). This must inevitably curb those flights of design fancy that would take a project into higher price brackets, though there is theoretically no need for exceptional design to cost above average. It is of especial importance to Pentagon that the engineering requirements for the services are fully incorporated at all stages, so that for example minimum duct runs and minimum heights can be achieved, as well as other more visible requirements.

Notes

1. 'Space planning' is a term for the work associated with layouts of interior partitions and furniture, especially associated with offices. It depends on systematic information gathering and analysis, and understanding organizational structure and consequent needs. It covers both more and less than would normally be encompased by 'interior design'. It originated in the USA, and DEGW were influential in bringing it to the UK and Europe.

2. 'Building appraisal' is a term used in DEGW to cover an extension of the space planning approach, which was developed by them out of the work on a multi-client study on the impact of IT on office building (*Orbit 1*, 1984). Buildings are analysed according to their capacity for use in particular ways, and assessed with regard to their suitability for different types of user.

3. Duffy, F., Cave, C. and Worthington, J. *Planning Office Space*, London, Architectural Press, 1976.

4. *Orbit 1*, DEGW, 1984.

5. White, W. *The Spirit of BDP*, Building Design Partnership, 1987.

6. *The Essence of BDP* (BDP Handbook), 1981, written to encapsulate the 'first 20 years'.

7. *Strategic Study of the Profession*, London, RIBA, 1992.

8. *The Essence of BDP*, Building Design Partnership handbook, 1981.

9. Ibid.

Part Four

The Range
of Work

9

A selection of projects

Variations of approach

Of obvious interest is the effect of the changes in architectural practice which we have described on the built environment which has been produced. Broadly speaking these might be expected to be of two kinds: a greater variety of new building types and sizes and a less consistent visual language for the products of architectural expertise. Whether these kinds of difference in building production will be seen as having led to an overall improvement or to a deterioration in the quality of our surroundings is perhaps an issue for wider debate than is appropriate here. That there is some basis to be found in the conditions of practice for changes in subjectively perceived quality is the point that can and should be made. What is built is not entirely independent of how it is designed.

A selection of the projects reviewed during our study are included here to help explore this relationship.[1] As with the case studies of firms, the reports deal with the state of these projects in 1991 and in some instances there have been further developments since then. The first two cover different relationships of a firm with a client over the whole of the study period:

- a traditional high-quality building design service: Templeton College (ABK)
- an innovative range of consultancy in interior design, space planning and strategic planning: for IBM (DEGW).

At the start of the period clients looked to architects to help define the fundamentals behind the brief for buildings of types that they would rarely have had the opportunity to commission. Today it may be that clients neither expect to learn so much from their architects, nor always wish to give them such full control over projects as used to be the case. The work with IBM illustrates aspects of the consultancy and user guidance work done by DEGW, and how these interact with more familiar, traditional design work. It indicates the complex structure of skills involved. It shows that these new skills have been around for quite some time, being refined and integrated into the world of architects. In many instances skills which were needed in a rudimentary form as part of the comprehensive service given in projects like Templeton College have been formalized and turned into special services for clients such as IBM. The second pair of projects can also be contrasted with each other as they represent commercial office projects from the earlier and the later parts of the period. The pairing illustrates:

- a large headquarters building on a cleared site: Halifax Building Society (BDP)
- speculative rebuilding behind an existing historic façade: offices in John Dalton Street (Stephenson Architecture).

The first is for offices for owner occupation by a commercial organization, needing prestigious headquarters, working in a receptive town planning climate and able to commission a state of the art building. The second is a complex urban renewal project where all the architect's ingenuity, both in design and in negotiation for permissions, was needed to give it a real chance of being built. Two further projects show aspects of specialization and the exploitation of a special skill in a niche market:

- energy conscious design: Yateley Primary School (Hampshire County Architects Department)
- historic preservation: Tudor House (Roger Mears Architects).

In the school, the personal interest of the job architect in pursuing the subject of energy management led to the timely incorporation of an aspect of design rapidly becoming more essential. It could not have taken place in the earlier phase of Hampshire's design endeavour, where repetitive projects were being carried out to fill the urgent need for schools. In Tudor House Roger Mears developed his personal specialism in faithful restoration of period architecture, thus establishing a potential niche for his practice. Finally, we include a project which epitomizes the changes in procurement styles:

- design subcontracting: stations for Docklands Light Railway (ABK).

This sophisticated infrastructure development work was undertaken for engineering main consultants by the same firm as was responsible for the first of these seven projects. The dedication to design quality seems unchanged but the legal and financial context in which it was applied has altered fundamentally. The format for collecting information about the projects is summarized in Appendix IV.

Templeton College Oxford (1965–90)

This is a project with a life virtually as long as that of the practice, ABK. It has taken place in distinct phases, seven in all, starting with a master 'diagram' for the planning of a business school, the Oxford Centre of Management Studies, on the edge of Oxford just off the Abingdon Road. It has continued, from time to time, with the construction of various phases of the college, as it has now become, and even with adaptation of the buildings designed first, to meet a new definition of their need. The work came to Richard Burton through his contact with the Centre's first Director, who had a similar role in another college for whom Richard was working. The Director was anxious to encourage the provision of buildings of high architectural quality in academic Oxford, and was convinced that in ABK he would find the appropriate level. The first phase of the work was opened in 1969, the seventh in 1990 (Figure 9.1).

In general the phases have all been fairly small and of relatively short duration, each completed within one or two years. They have been carried out by the same local builder in all cases except the first phase, and with a small office team of Richard Burton and one job architect, who changed as the years went by. There were also changes in the client team, sometimes disturbing the harmony between architect and client/user. The strength of the original concepts and architecture, and the close involvement over such a long period of growth and development of the college, have made links that all are reluctant to break, and great efforts have been made to maintain continuity of design input by ABK.

Early discussions between Richard Burton and the college Director about management education and its role were an important foundation for the development of the brief. The client's objective was to achieve a plan that would respond as the future unfolded and a role for management education became more clearly defined. He did, however, have some clear ideas of starting points to which the initial building plan was to correspond. The original matrix on which the plans are based was generated out of the general and specific requirements of the brief, and owes a lot to the combined discussions of all three partners at ABK.

157

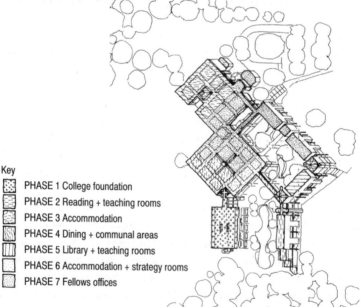

Key

🔲 PHASE 1 College foundation
🔲 PHASE 2 Reading + teaching rooms
🔲 PHASE 3 Accommodation
🔲 PHASE 4 Dining + communal areas
🔲 PHASE 5 Library + teaching rooms
🔲 PHASE 6 Accommodation + strategy rooms
🔲 PHASE 7 Fellows offices

Figure 9.1: Templeton College: A master plan concept with a strong planning structure allowing growth. The first phase set up the elements, and six subsequent phases over 20 years have allowed the expansion originally required in the brief. © Ahrends Burton and Koralek.

Since the creation of that framework, the projects, which have been under Richard's control, have been designed and directed by him, with the job architect contributing to resolution of details of arrangements and construction. On projects of this scale the manual production methods that were used in the early days have not been replaced by CAD. The projects have been managed by regular weekly site visits, and have had the benefit of continuity from the construction team.

None the less the office has found that this scale of project cannot be run very economically. The commitment to the client and to the early project are an essential part of what ABK sees as its professional role. However, the amount of attention required for a smaller job is disproportionately high compared with a larger one, and as economic constraints become tougher, and a business ethic jostles the professional one, the challenge becomes how to be able to afford to offer the professional service to a long-term client who should not be abandoned, and to continue to work with a design idea whose adaptability it is still desirable to test.

IBM (1971–present)

As Frank Duffy said when talking about IBM, there has probably never been a day, let alone a long period, when DEGW has not been working with and for IBM in some way. It is common for architects to carry out several or even many projects for the same client. In this case the involvement of DEGW with work for IBM has been in a wide variety of types and sizes of projects, executed in a wide range of circumstances and locations. Over the period 1971–88, 75 reports were prepared for IBM from the different projects. Both the client and DEGW have developed together in many respects, gaining insights, sharing techniques, and understanding objectives over twenty years of interaction. The general overview of the history of work with IBM is therefore helpful in understanding the development of DEGW and their particular way of working.

The original contact with IBM predates DEGW, as a project was done by JFN for IBM Netherlands in their Amsterdam typewriter manufacturing plant, 1971–75. Luigi Giffone principally worked on the studies for this job, which included some space planning and interior design work, as well as a substantial graphics project. In 1972 JFN again took on a project, this time simply space planning, for IBM Italy at Segrate in Milan. For this work the team was Frank Duffy, Luigi Giffone, Luigi V. Mangano, and Colin Cave. Here a very important core of the future DEGW can be seen in embryo. Out of these contacts, after the formation of Duffy Lange Giffone Worthington, a seminal project from DEGW's point of view was undertaken. Known as the 'Open Plan Study', it was a comparative study, looking at IBM offices in four different European countries, and was in the form of strategic guidance about what IBM should look for in performance from their workplaces. A very early project of this type done by JFN, in which DEGW has developed great strength since, was the 'Space Standards Studies' for IBM UK in 1972–3. Rather than simply taking a particular building and creating a design solution to fill a particular need, the approach analyses at a more general level, often across a range of buildings, the basis and rationale for design decisions, thereby creating a framework for further decisions.

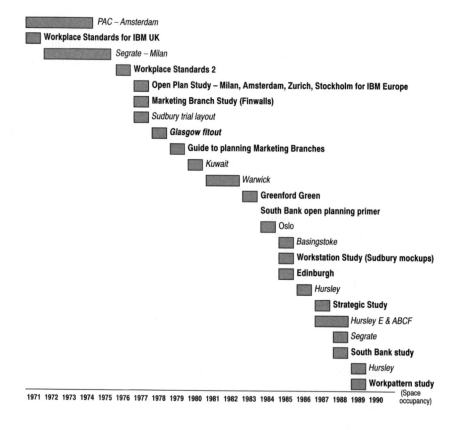

Figure 9.2: A succession of projects carried out by DEGW (initially as JFN). The job names in italics indicate space planning and interior design projects, the job titles in bold indicate projects reporting on strategic guidelines and development of standards. A long relationship with IBM has been one of the corner stones of the development of the practice.

The pattern of movement of senior staff within IBM, an international giant, has been significant in spreading the knowledge and influence of DEGW's work. In the UK there have been some vital consistent client contacts in the central property department, and in the property department concerned with the London offices, which has maintained the continuity of understanding that makes it possible to build usefully on previous experience. There have also been important IBM staff contacts who have been involved in one project and then later been moved by IBM to another part of the world.

This has presented an opportunity for further international contacts and work, and been one of the ways in which DEGW has developed its strength in working in different parts of the world, both on a rapid strategic consultancy basis, and when needed to form a local project team to do a larger physical job. Examples include work done in Kuwait in the early 1980s, which was commissioned by IBM Europe, through the client contact involved in the Amsterdam work and in the 'Open Plan Study', and the further work now being done in Milan for IBM Italy at Segrate, by DEGW's Milan office.

The type and amount of work done with IBM over the last twenty years has fluctuated with the economy. The period 1971–3 was busy, but the energy crisis curtailed work during 1973–6. From 1976 to 1981 there were a number of projects covering the wide range of different types of work that DEGW have done for IBM, such as space planning in Sudbury, the marketing branch study in a number of UK branches which led to the development of the fin wall, and the strategic guidance work on planning marketing branches that developed out of this. Another lull followed, although work did not just come to an end, as some larger projects still needed to be completed, such as space planning at Warwick, and other smaller strategic guidance studies took place. As the economy picked up in the mid 1980s so did DEGW's work with IBM, who needed to expand in the Basingstoke area, and took on space in one of the Gateway buildings. This project was a product of several earlier connections, as DEGW knew the building from an earlier project, space planning it for Wiggins Teape, the previous occupants. A busy period followed with a number of large projects, in space planning and strategic studies, at Hursley and at Southbank among others. As the decade ended and another recession loomed, large projects again slowed down. Strategic discussion of workplace issues, however, remains a live activity between IBM and DEGW who have joined in an active forum with both IBM Europe and the new IBM property company, Procord, as well as with other commercial organizations.

The complex set of interwoven relationships with connections to a variety of other organizations as well, and the evolution of ideas as the years have passed, have made DEGW's work for IBM of critical importance to the development of the practice. There have been examples of most of the types of work that the practice undertakes, with the exception of a new building, since the site and strategic planning work has generally been done as background for design work by other firms. Many of the projects have been undertaken by one or other of the group who worked together on Segrate in the early 1970s. Over time other members of the office, specializing in space planning or in interior design, have taken on leading roles in some of the projects. In some cases the social psychologists at BUS or the furniture and graphic designers in T3 have played a larger role.

Halifax Building Society HQ (1968–73)

BDP's initial contact was through another local client. The Pro-Vice-Chancellor at Bradford University, a long-standing and extant client, was a director of the Halifax. The building was to be a new headquarters, with all the necessary staff accommodation and storage for an expanding and success-ful company. The final cost was £7.5 million for 350,000 ft^2. It was commissioned in 1968 and finished in 1973.

There was a long briefing period, of about a year, in which time the client and the BDP design staff travelled together to see HQs in Europe and the USA. They were particularly influenced by Osram HQ in Europe and by Connecticut Life HQ on the east coast of the USA, an out of town site. The need had been established early for a few large floors, not a stack of smaller ones. This building was a product of the time when designers, and probably their clients, thought that an organization could be sufficiently defined to get the 'right' building. There was extensive analysis of the paper factory aspect, and the need for secure storage of what was predicted to be 25 per cent of the title deeds of UK house owners. As a result there was a storage warehouse below, with clerical workers above on an open plan floor fed from the warehouse. There was also a great deal of thought for staff welfare and amenity; the Halifax was a paternalistic company with enough money to fund a good building as it was expanding at a phenomenal rate. At the end of the long briefing process an exploratory design was produced, and immediately accepted. No iteration was needed.

The contract was let in two stages, a schedule of rates was prepared and contractors tendered against a set of general arrangement drawings. BDP wanted the contractor on board before detailed design was done in order to get input on buildability and speed. This was needed as the Halifax's business was expanding fast and the storage space was urgently needed. The client's desire for quality, and willingness to commit both cash and their own time,

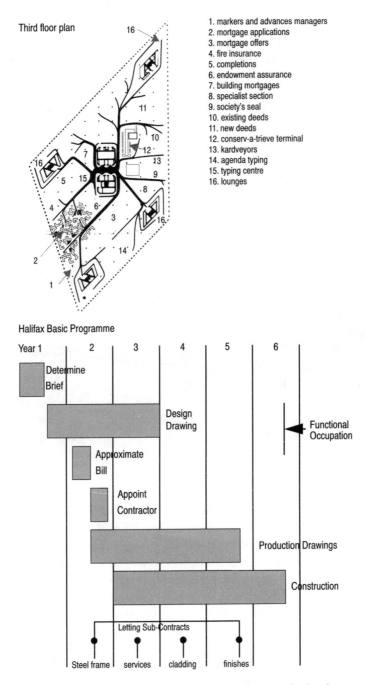

Third floor plan

1. markers and advances managers
2. mortgage applications
3. mortgage offers
4. fire insurance
5. completions
6. endowment assurance
7. building mortgages
8. specialist section
9. society's seal
10. existing deeds
11. new deeds
12. conserv-a-trieve terminal
13. kardveyors
14. agenda typing
15. typing centre
16. lounges

Halifax Basic Programme

Figure 9.3: The Halifax headquarters required complex co-ordination for an elaborate set of systems, both spatial and servicing.

163

made it possible to have a 'teach-in', explaining the project and its commitment to quality, for each group of new trades that came on stream whenever there was a major change on site.

The BDP team was about 40–50 strong, relatively consistent in size for about two years. There were about fifteen architects most of the time, plus structural and services engineers and QSs. Some engineers worked simultaneously on other projects but the architects and QSs were largely involved in this one large project which was very sophisticated and ambitious for its day. Some people moved in and out of the team during the course of the project but the core team was fairly stable.

The design group present at the briefing stage included six architects who were still there at the end of the job. The concepts were largely generated by Bill Pearson, the partner in charge, but were then developed by the design team and each person brought their own ideas to this process.

Figure 9.4: Halifax Building Society headquarters.

There was a regular sieving of the design solutions arrived at by a combined group, including the client, the contractor and BDP. This group oversaw the way in which solutions met the brief of use and cost. The team was organized to work on different packages, but each designer was given responsibility both for an element and for an entire floor and the bringing together there of all the different elements. The job architect was an architect with a management type of specialism, though this was not stressed at the time as it was not yet a 'respectable' career choice for a designer. The engineering was very sophisticated for its day, and threaded services through the massive structure. The cost plan was able to accommodate the great expense of the sophisticated structure and services.

Drawings were all done on tracing paper, as CAD had not yet been introduced. This project was used as a trial for introducing the standard reference system for construction (Sfb), and an enormous amount of paper was generated by this process as the team was over-ambitious about the degree of detailed distinctions made. The results of this experiment were fed back to the practice and used to modify the way in which BDP used Sfb afterwards.

John Dalton Street, new offices (1988–93)

This project has been a challenge for Stephenson's practice as it is for a new office building to be constructed behind a retained façade in a narrow city street, for a contract sum of £6.2 million (Figure 9.5). Originally a client of long standing owned the property and planned to refurbish it. This was a potential project for the office in 1988. The client, however, became less active in the development world, and sold his interest to Barlows. The entire project went quiet until late in 1988, around the time of the split-up of the partnership and the formation of Stephenson Architecture. The client was now a group called the John Dalton Partnership, formed from Barlows joined by the Church Commissioners. The latter are active in the development field with around 50 projects in the UK as well as some in the USA. Roger Stephenson was asked for a diagrammatic study of the site, and the new clients, who were looking towards a new-build project, were convinced that the practice could create the hybrid building that they needed. A fairly elaborate and high-quality Victorian façade, a relatively central city site, and a listed neighbour, contributed to the necessity to retain the façade, while the cost equation required the creation of economic and high-quality offices behind it.

The project received planning approval in early 1990. Meeting the requirements for the façade retention, and the connection of the new building to the existing one, required considerable negotiation. Work was started on site in April 1991. The contract period was planned to be 82 weeks so that the building would be finished by late 1992. The JCT 80 contract was amended by the clients to suit their requirements. The site is adjacent to a listed Catholic church, and that fact, as well as the narrow city street on to which the building fronts, mean that the contract has had to lay down considerable restrictions on the possible working times for the contractors when using noisy processes, or dealing with deliveries and rubbish removals.

Initially, associate Paul Iddon, who is in charge of the job, having taken it over from Stephenson's sketch diagram stage, was virtually on his own as a one-man team. The first seven months were slow, being spent on listed building and planning negotiations and consents, but a project of this size, once under way, has required a sizeable team for some months. There have been four qualified architects as well as technicians and interior designers, in a team of nine to ten people for a six-month period. In the early stages some project management advice was taken from a firm with whom the practice is familiar from other projects. The team is working with a group of structural engineers from Yorke Rosenberg Mardall (YRM). As it has been a comparatively large project for the practice, there has sometimes been considerable workload pressure on the team, who have had to prepare a large quantity of drawings.

Figure 9.5: Section and elevation of the John Dalton Street project.

The brief was jointly developed by the practice and the client, who as developers were concerned not with a specific user brief, but to achieve a building that would attract a range of possible users. The design was largely developed by Paul, based on Roger's initial sketch, with some additional occasional input from Roger and others. Much of the design process has been carried out with the help of models. Seven have been made as the design progressed, of which two were presented to the client for clarification, and the others were used as a design tool. The principles around which the design is based include seeking to use traditional detailing, with 'no gymnastics'. There was continuous feedback from quantity surveyors on the need to reduce costs, and the designers were put under some pressure to reduce the size of the atrium. As an approach to cost containment through the course of the design, the principle has been established to retain the use of high-quality finishes, but where it has proved necessary to make savings, to reduce the areas to which they are applied rather than the material specification. In addition rules have been established about which aspects of the design are sacrosanct and cannot be tampered with as part of a cost-saving exercise. Of particular importance are common areas such as the entrance details, the atrium base, and the way in which the link is formed between the existing façade and the new building.

This project has also been the one on which the practice has developed its CAD capabilities. The main body of the drawings has been done on CAD, and there are two people who are CAD literate and one person using the equipment full-time. Over 100 drawings were needed for tender and altogether over 300 have been prepared, mostly on CAD, though one-off drawings and some details have been prepared manually. The larger scale of the project and the fact that CAD was new to them meant that they were learning on both fronts simultaneously, and this led them to underestimate the time it would take to prepare the drawing packages. However now, over two years and with some pain, they have developed an expertise in this area which needs a steady diet of similar projects to exploit it.

Management of a larger project and a team that for some of the time has been about one-third of the office staff has been an issue in this project. It is harder to be democratic and at the same time 'speak with one voice' on the issues of both design and production. In the office the team are dispersed into different locations on the office floor, with not more than three sitting close together. None the less there is pressure from the team to bring them together as they feel they need to be a coherent group. They also feel that in the course of the project they have discovered that they could do the project management themselves.

Yateley Primary School (1977–79)

The Yateley Newlands Primary School was one of several schools being commissioned for the Hampshire Education Department at the time. The project was under the control of Mervyn Perkins, a young architect in the department, who on looking back can see that its development was part of a series of changes that were taking place generally in the department, though at

167

the time this was less apparent. The project was completed in 1979 and since then has won a number of design awards and, more importantly, has proved a popular and successful building for the users.

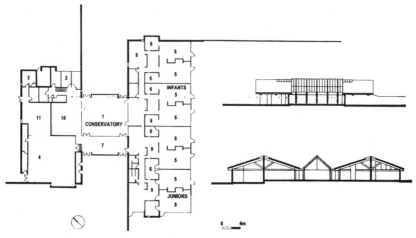

1 RECEPTION 2 OFFICE 3 STAFF 4 HALL 5 CLASSROOM 6 CLASS BASE/TUTORIAL 7 RESOURCE 8 LIBRARY 9 SHARED 10 MUSIC AND DRAMA 11 KITCHEN AND SERVERY

Figure 9.6: Yateley Newlands School combines the educational benefits of a deep-plan layout with a selective approach to environmental control using daylight, natural ventilation and solar gain.

Figure 9.7: Photo of atrium/swimming pool.

The background of the job architect in this instance is of considerable importance in the development of the project. His total involvement in the project, as a young and relatively inexperienced member of the department, is equally a reflection of the management style practised by Colin Stansfield Smith in his drive to achieve high-quality design.

Mervyn Perkins, when a student at Portsmouth, had spent his 'year out' at Hampshire in 1973. His experience, even before Stansfield Smith came to the department, happened to be on an educational project that was not SCOLA, a sixth-form college, under a job architect somewhat sceptical of SCOLA, and with an inclination to treat projects as one-off. After qualifying he returned to Hampshire as one of the first of the new chief's new appointments, with no expectations of using a standardized design kit to answer problems.

The energy crisis in 1973 had generated concern about energy in buildings, and the Architects Department sponsored Mervyn Perkins for a special research project looking at energy use in schools. Prima facie the SCOLA schools should have had comparable energy use when corrected for environmental context. Ten *identical* schools (there were 23 built of a single design in the county at the height of the schools programme) were looked at, and differences were found that could not be explained by physical parameters. With a closer look, putting twenty fifth-year students into four schools for a week, and looking at 32 classrooms, the project came up with even wider variations between the classrooms, 100 per cent in some cases, than between buildings. This work was being carried out during the period that Yateley was being designed and had a considerable influence on Perkins, who had a part- time role in the research project. It taught him the importance of thinking about a school as being about a teacher and a class living in an environment, so that any energy control systems depend on how 'user friendly' (not a term of that period) they are for their success.

The brief for the school was prepared by the Education Department's primary adviser Jock Killick. As he was an ex-teacher the brief reflected his own personal statement, with which other teachers might not necessarily agree. The brief left many aspects of the design open, for example not prescribing the degree of enclosure or openness required. The head teacher was appointed three months before the end of the design period and thus could also play a role. However it was one of a number of briefs which were fundamentally rather similar, and which over this period of change in the architecture department were able to act as the basis for a very wide variety of designs for identical needs, but in different physical situations and in the hands of different designers. This formed a direct contrast – even, on one view – an extreme reaction, to the SCOLA period.

The project came to the department in 1977, at the time when departmental reorganization into geographical areas was going on. A site for two schools was looked at with the county Estates Department, and a feasibility study for one was decided on. For a year after that the project went quiet, and then design was started in late 1978. The design process was treated in a similar way to a student project, being criticized, honed up and championed by the Chief Architect. Colin Stansfield Smith presented projects at committee to the council so he became involved before they went forward. There were three or four sessions to develop the design, not necessarily at the drawing board, with Stansfield Smith and his deputy John Robinson, though after approval by committee there was less involvement from the senior staff.

The detailed design and production drawing phase was undertaken rapidly, and by a team of one, Mervyn Perkins. The committee approved the scheme in November 1978 and it went out to tender in January 1979, Perkins having done all the 40–50 drawings himself. At the tender stage it went over budget, so alterations were required to save £40,000. This was achieved in the ten days before going on site, keeping the design within the same foundations, but for example reducing the atrium, altering stanchion bases but not changing reinforcement, etc. The building went on site in May 1979, and there was a two-phase handover in November 1979. This was therefore a very rapid project, using one full-time architect in the department for a year.

There were also other players in the team. The QS work was done in-house until after approval by committee and then went to an outside company. The structural engineers, a well known firm, had been used on other projects by the department. The department liked to get engineers in on projects at an early stage in the design, and had early working sessions to exchange ideas and get preliminary designs well grounded. The choice of outside consultants tended to come from within each directorate, based on their particular experience. At this stage, before Hampshire became well respected for design skills, nationally known consultancy firms could be expected to give variable service, though now that the department have such credibility as designers they are more likely to be able to command consistently better treatment from outside consultants. The conservatory detailing and components were provided by a small organization who had worked with Norman Foster on the Willis Faber building in Ipswich and a single tender was used. The components were made in Ipswich and brought to site in a Transit van.

Tudor House, Chelsea (1986–92)

This project was a large one for Mears, and has been very important for the practice. In 1986 it came to Roger through a personal contact that he had made as a result of taking the SPAB course. The house was built in 1717 in Chelsea and is listed grade II*. Over time many alterations, minor and major, have taken place, and there has been insensitive maintenance work done.

Initially there were delays while the client decided definitely to carry out the necessary work, and during the course of the project the building has in fact changed hands, with the restoration project as part of the commitment of the new owner. The project has been a slow one, done with meticulous care, and used by the SPAB as a case example for 'how it should be done' for a recent course. The work has included repairs to the roof and to the brick and stone; structural repair, especially to ensure the continued viability of past structural alterations; and the introduction of heating with humidity control. The house has been brought up to modern standards where this is possible while remaining in sympathy with the original building.

Work started on site in January 1991 and was expected to finish in October 1992. It has both enabled the practice to acquire an unassailable 'track record' in this kind of work, and provided adequate work to keep them busy and funded in an otherwise lean time for most architects. It has reinforced the desirability and possibility of creating common ground with other architectural firms who specialize in this type of work, one of which was the source of the original contact with the QSs for this and subsequent work. In order to arrive at a price for the work there was a painstaking process of examining every element of the building in the company of a specially skilled QS. This resulted in detailed specifications for all work required, never becoming general by the use of terms such as 'repair as required', but describing in detail, for example, which members in a particular sash window assembly would be replaced, and which retained or reinforced.

Though the job is larger than is normal for the intermediate form of contract, this level of detail has made it possible to use the contract as virtually all the necessary work has been described and measured and could be priced by the contractor. The original work in assembling the specification was done largely by an assistant specially hired because of his skills and training with historic building work. He left when the project went into temporary abeyance, but when it was revived another architect was found with a similar specialism and a sympathy for the approach to renovation. This approach retains as much as possible of the existing fabric, replacing only those parts which are 'worn beyond all possibility of repair'. Had the full scale of this project been clear at

Figure 9.8: The specialized niche dealing with work on historic buildings requires not only architectural skills but also access to skilled crafts.

the outset, it is possible that Roger would have hired a slightly larger team to enable it to progress faster. However, it would still have had the hiatus due to the client's indecision, and the slower progress has, Roger believes, led to a better product as a result, partly because he has been able to learn more during the course of it than a faster project would have allowed.

Docklands Light Railway, twelve stations (1987–present)

This project is different in several ways from the small repeated series represented by Templeton College also undertaken by ABK. It is much larger, took less time overall, and was carried out under totally different contractual circumstances.

The project was to design the twelve stations for the second phase of the Docklands Light Railway (DLR), across the northern edge of the Royal Victoria Dock area, as far as Beckton, in London's East End.[2] A transport project in that location, at the height of the property boom in the late 1980s, it has characteristics typical of its period. The way in which it was carried out, on the basis of an initial fee bid, and in the second phase with the architects as subcontractors to the lead consultants, the engineers Maunsell, is also characteristic of recent years, but would have been unlikely earlier in ABK's history.

The appointment was in two stages. The engineers Maunsell, had already been appointed by the DLR, and preliminary work had been done on the station requirements by their team, which included architectural staff. The London Docklands Development Corporation (LDDC), the ultimate client for the project as a whole, wished to broaden the possibilities for the detailed solution stage and approached several design firms to ask for tenders for twelve stations. The submission, a fee bid, was to be for a kit of parts that could be used for all of the stations, combined with a contextual study of the different locations and their specific requirements. ABK won this stage with their first fee bid proposal.

They started with an analysis of the first phase of the DLR, consisting of fifteen stations designed by Arup Associates and executed through a design and build contract. They used this part of the work to identify issues that were fundamental to the needs of the DLR and the LDDC, and to form the basis for the brief to which the new stations would be designed. The earlier phase of the railway was important. Design continuity was needed, combined with a high-profile image suited to the 'red hot' development situation in Docklands at the time, where schemes were chasing fast upon each other's heels, and the recession had not crossed people's minds. The report back on the earlier

173

phase created the necessary dialogue from which a brief could be derived. On the basis of this, design proceeded and took the project forward.

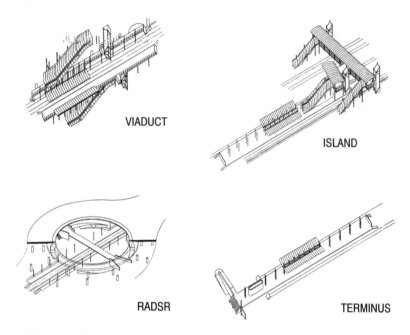

Figure 9.9: The stations were designed to allow a 'kit of parts' to be assembled to suit the configuration of the site: products as well as project design.

This early design stage took place in a very short time frame accompanied by a large number of meetings. There were fortnightly cycles for three different groups, some of them consisting of twenty people or more, as well as one-off miscellaneous meetings with specific interested parties. This intensive period of communications lasted about three months, between autumn 1987 and spring 1988. The ABK team consisted of two separate groups of four people each, one for the development of the kit of parts, and one for the context studies, where each member of the team considered three stations.

ABK were asked to take the project to tender stage on a new footing, carrying out only design, detailing, specifying and drawing, as a subcontractor to the lead consultants, the engineers Maunsell. While not accustomed to this role, they accepted it as appropriate in this context, and carried it out to the satisfaction of all concerned. When the contract with Maunsell was signed – in September 1988 – a very clear and full statement of what drawings Maunsell would require, and ABK would prepare, was set out, and formed part of the contract, to avoid the possibility of later confusion. This was a necessary feature of a subcontractor relationship of this kind. There was no option in this project for ABK to work in the more traditional role of lead consultant, and the experience of setting out the relationship and obligations when in the role of subcontractor was a new one. Considerable effort went into clarification, legal advice was sought on the form of contract, and the experience gained has since proved useful, as other projects with similar relationships have been undertaken.

The production drawing phase for the main contract and additions, such as the two bridges that were commissioned separately, was complex. It took about six months and involved numerous consultations, for example with the Disabled Advice Group used by London Transport, with the Railway Inspector, and with representatives of the different areas in which the stations are located, to ensure all relevant requirements were met. During this period, the lead consultants Maunsell continued to use architectural staff in Maunsell Associates to bring together all the information from ABK, and from other specialists in their team, into general arrangement drawings (GAs). ABK provided the material required for the specifications, and the drawings for their part of the project, and Maunsell put together the tender documents for the end of 1988. At the start of the project CAD was not fully integrated into ABK's work, so the drawings were done by hand. The repetitive nature of the kit, and the importance of the large-radius curves in the route affecting the station forms, could have made it a good candidate for CAD. However the GAs, where the greatest benefit of CAD was to be had, were, as mentioned, not prepared by ABK.

Large projects use a high proportion of staff in a medium-sized office such as this, and most of the team for this job were hired specifically to work on it, though a few had worked with ABK before. The project was under the direction of Peter Ahrends and the team was managed by associate David Cruse. Partner involvement remained close throughout, which is normal in ABK. Design ideas at all stages are the subject of discussion between the partner and the team members. Senior-level involvement in detail is seen as essential in order to keep quality high. Careful recruitment of the team is seen as crucial to good execution.

The stations are on site, nearing completion, and there is one staff member remaining on site and one in the office. It has been a new but worthwhile experience. Several fee bid submissions have followed this one, and further work has been sought in similar subcontractor relationships. The experience with Maunsell was a good one. The teams got on well both personally and professionally, and the success of this project has helped ABK to change its way of working when a project requires it, more easily than might otherwise have been the case.

175

Notes

1. The full number can be read in Eley and Symes, *Recent Changes in Architectural Practice.*

2. The number of stations was altered during the contract as a consequence of route changes.

Part Five

Questions for the Future

10

Conclusions

The response to change

In this book we have reported on a number of studies of architectural practice, intending to show how this once highly traditional professional activity has responded to economic and social challenges. The designer's role has been at the mercy of dramatic forces for economic change in the construction industry. It has also been at the forefront of a long-running debate over the legitimacy of traditional procedures for making environmental choices. The research shows that the profession has at times responded robustly to these challenges. It also suggests that leading members remain very strongly committed to a mission and a vision which could easily have been abandoned. It also shows, however, that a significant proportion of the profession have felt poorly prepared to cope with the new situation. If this is not corrected the quality of the designed environment will surely suffer.

Thus a contrast between the first and last of the seven projects illustrated in the previous chapter suggests that key architectural practices have adapted well and are alive and kicking. The case studies show that great changes have been experienced and many new options explored. The survey results indicate that a good proportion of principals or partners feel that their ability to face the future is none the less still unclear. Various questions were raised in the first chapter about options for the future and it is now time for us to return to them.

Questions addressed by the research

The first question raised concerned the development of specializations within the profession. Here the evidence is ambiguous. The case studies show that some, carefully selected, firms have certainly become more specialized and can earn their keep in this way. A conventional wisdom used to be that this would be most likely to happen within very large organizations, but the case study of Mears shows that specialization may in fact be an excellent survival strategy for the very small firm. And perhaps it is worth pointing out that there are a very large number of very small firms in architecture and, strictly speaking, no very big ones. Even BDP, which is much larger than any other architectural firm, does not compare for size with consultancies in other professions.

On the other hand the questionnaire survey shows no clear pattern of specialization among the principals of private sector architectural firms. Broadly reviewed, they pretty well all consider their major activity is designing or, which for senior people is much the same thing, managing design. Some businesses respond to market opportunities by taking on more of one building type or of one section of the design process than of others, but very few believe that this gives a fundamentally different structure to their work. The projects illustrated in the book suggest more emphasis in recent work on infill than on new build and on staged building programmes than on grand set pieces, but these observations may owe as much to short-term changes in the economic and political context as to fundamental alterations in what Mill would have called the division of professional labour.

The surprise is perhaps that so much of the average architectural practitioner's time is spent on the *management* of the design process, no matter how much of a niche he or she works in. One strong possibility for the future must therefore be that *this* will become a real area of specialization and produce a new and enduring sub-profession. In manufacturing industry, design management is seen as a specialized function to which graduates can move during their careers, and it may be that architectural design management will be seen in this way as well in future.

The second question in Chapter 1 referred to the introduction of new technology. Interestingly one of the small firms interviewed suggested that the design management role could not have been retained if computer-based systems had not been put in place. Pressure has clearly been felt in this area and partly solved by mechanization. Of course there are strong processes of expectation and emulation which tend to mean that if leading members of a profession adopt certain procedures, sooner or later most others will feel they need to as well. So there is a sense in which computerization has arrived because it has arrived. But it is none the less real for that. The consequences of its arrival certainly deserve discussion.

One consequence may be that fewer jobs and less permanent jobs may be available with a given workload. Not only do computers work quickly, they also hold information and keep records of decisions. In many types of design, the need to research a problem before considering solutions may diminish rapidly. Again, the importance for a firm of the experience that their employees may have had would evaporate rapidly if that experience could be stored in a suitably structured automatic memory. Even negotiation between designers with different professional specialities, a function certainly demanding expert knowledge, can become redundant if machines can talk to each other. Pressures for lessening long-term commitments to highly skilled employees will be accentuated by the growth of costs attributable to constantly upgrading equipment in a competitive economy. The likelihood thus is that design – for long a true cottage industry – will soon experience a quite radical substitution of capital for labour. Interestingly this does not *necessarily* mean that only large firms will survive. It seems equally possible, to us at least, that small firms will proliferate, providing they can be adequately capitalized.

A third question was asked in Chapter 1 about the growth of large private firms, and a fourth about the decline of the public sector. Because of the way the questionnaire was set up, our case studies have told us most of what we have learnt about this double development. There seems to be a consensus that although much useful knowledge has been dispersed and that organizing its updating is an increasingly complex task, private firms like DEGW are in fact engaged on it and may do it very well. On the other hand there is a hint in the questionnaire response that architects at the beginning and the end of their career are less interested in pursuing the contextual questions which surround building than those in their middle years. If this is so, a paucity of public employment in architecture could be reducing opportunities for them to pursue this interest at what is, presumably, a rather productive stage in their careers.

The fifth question concerned the organization of the professional body, and this led us to include in the survey a long sequence of questions on perceptions of the educational system. Practitioners were asked to enumerate in considerable detail the areas of work for which they felt well or poorly prepared. In the light of the comments just made about the significance of management activities for our respondents, it will come as no surprise if an emphasis is also seen to have emerged on the need for management education.

To recapitulate on the findings of Chapter 4. In the general overview of responses it was stated that:

> ...*we quickly see areas in which the principals report an architect needs training but which are rarely a significant portion of the architecture curriculum. While schools usually address building technology and history of architecture in depth, the other subjects receive minimal or simply no coverage. It would appear that computer-aided design, office management, budget management, construction management, human*

181

> behaviour, marketing, research and accounting as well as many other
> areas of expertise are needed by architects... As discussed earlier, these
> data provide additional confirmation of the importance of management
> skills both in the profession of architecture and in educating
> practitioners.

Later in Chapter 5, we continue:

> ...it is the central age group which is most likely to report an architect
> needs training in [these topics]... they are most likely to report that an
> architect needs training in areas... which are usually not part of an
> architect's education.

There was no significant difference of attitudes expressed on this matter by the
principals of large or small firms, or of multi-disciplinary ones, but, again from
Chapter 5:

> ...principals of firms which specialize in... commercial or industrial
> projects or institutional or public projects are most likely to report that
> budget management, office management, project management and
> construction management should be part of an architect's training.

And yet we know that this is rarely the case. And of course undergraduate
architecture courses are already overfull with studies which in their own right
are valuable and valued. Perhaps we should argue that what is needed is a
strong programme of life-long learning, directed at private practice architects
in mid-career, working on complex building tasks for sophisticated client
groups, in all aspects of modern design management techniques.

To look at the situation from a wider perspective, the evidence of the studies
reported here is not altogether clear on the degree of re-education which has
taken place. We interpolate however a strong possibility that the transformation of design practice which certainly has occurred *was insufficiently
supported by suitable training programmes*. Although the small number of
leading professionals we talked to in the case studies have seen that new
knowledge was being used by others in the field and gained access to it
themselves, most of the large number of ordinary practitioners who answered
our questionnaire were not apparently able, at least as yet, to update
themselves. If it is partly the function of the professional body to see that they
are able to do so, that body can hardly complain if its legitimacy as an
exclusive expert body, or its members' prerogative to undertake much of the
work they would like to see in their portfolios, has been put in question.

Interesting intellectual and practical problems follow from these observations. The solution of the practical one – how to determine ways in which continuing education of a suitable type can be developed and financed – is presumably the responsibility of the RIBA and of the educational institutions as well as of the practitioners. We hope they will progress thinking on this question as soon as they can.

Responsibility for solving the intellectual problem lies nearer home. Our findings suggest that although the profession still has a quasi-monopoly on expert knowledge, the terms of this 'social contract' are changing as restrictions on competition are removed. Architectural work is now more open to market forces: what seems to follow is that the architects must *gain more management skills* if they wish to regain their previous level of activity. Architects must find new clients, broaden their focus, not narrow it, as is often suggested, and learn how to *apply* their knowledge to a wider range of needs.

We think that they need a strong organization, such as the professional institute has sometimes been, to institutionalize (and help raise funds for) the development process which is required. How such feedback systems really work, and how appropriate professional institutions are to support them, are the kinds of question raised up to a century ago by sociologists like Durkheim and Weber. They are still worthy of study. So we make a case for further research, either by means of case studies investigating the adoption of new learning procedures by professional firms, or by a survey to find out when and how such firms adopt the training procedures they seem to require.

Our final question in Chapter 1 was to know whether architects remain personally enthusiastic for their work and committed to their profession. Despite, or perhaps because of, all the difficulties they have experienced in the last twenty years, the answer for most of them is undoubtedly yes. Not even for money would the respondents to our questionnaire give it all up and start again in another direction. They are hooked to a way of life. For them, surely, efforts to maintain it must be worthwhile. We hope that readers of this book who are not members of the profession also feel it has skill and expertise to offer to them.

Architects in the future

In a recent issue of an American journal the question was asked: 'Can the profession be saved?'[1] The author stated, 'The very idea of what constitutes architectural practice requires substantial expansion', and mentioned parallels with the engineering, medical and legal professions. Engineers 'depend less on rhetoric and more on the quantification of what they do'.

The medical professions 'have turned the general practitioner into a kind of coordinator of highly-paid specialists'. The enormous success of lawyers in developing for their discipline applications in a whole range of business, government and informal working situations is attributed to their 'unique form of synthetic analysis'.

No doubt architects have to learn from all three of these examples. If we had to choose one which gave the clearest indication of the way our profession is going, however, it would be the last. There are many, many places where architectural knowledge and skill can find an application. Good architectural design continues to be needed, everywhere. But the profession must develop the management skills needed to help it seize these opportunities. To do this rapidly it will need considerably more sophisticated educational institutions than it can currently rely upon. If architecture can move forward in these two ways, it will surely be able to continue to satisfy society as well as its individual members.

In a time of such rapid political and economic change as has been portrayed in this book, architects need to develop their own equivalent to 'the social learning which takes place in professional bodies like the the BMA and the Law Society'.[2] In our research into the state of the architectural profession we have become very much aware that 'The learning acquired in those organisations, which make up civil society and which mediate between the family and the state, tends to be ignored but merits serious study.'[3] We hope this book has made some contribution towards the fulfilment of that need.

Notes

1. Fisher, T. Can this Profession be Saved?, *Progressive Architecture*, February 1994, pp. 45–9, 84.

2. Economic and Social Research Council, *Research Specification for the ESRC Learning Society: Knowledge and Skills for Employment Programme*, ESRC, 1994, Section 3.8.

3. Ibid.

Appendix I

Two decades of change

Date	General context	Building Industry developments	Architectural events
1968	• Martin Luther King assassinated • Paris *événements* • Labour loses power	• Town Planning Act • Third London Airport Commission	• GLC plan for Covent Garden • Collapse of Ronan Point • *The Lord's Prayer*
1969	• Trial flight of Concorde • Maiden voyage of QE2	• Skeffington Committee	• Greater London Development Plan • Plan for Byker, R. Erskine • Shelter GIA in Liverpool • Monaco design, Archigram
1970	• Liverpool Docks and Harbour Board collapses	• Structure Plan for the South East	• Liverpool bans multi-storey housing • Milton Keynes Plan
1971	• Census • Decimal currency	• Buchanan dissents from Third London Airport Report	• Florey Building, Oxford, J. Stirling • Covent Garden Public Enquiry • *Air Structures*, Price, Newby and Svan • *Civilia*, de Wolfe
1972	• Murders at Munich Olympics	• Salaried architects' working group meets • Poulson investigation	• *Long Life, Loose Fit*, A. Gordon • *Adhocism*, Jencks and Silver • Reappraisal of Thamesmead • Covent Garden market moves to Nine Elms • Northampton TH competition won by Dixon and Jones • Lillington St, Darbourne and Darke • Demolition of Pruitt-Igoe in USA
1973	• Opec oil price quadrupled	• Sydney Opera House opens	• National Centre for Alternative Technology • *Defensible Space*, O. Newman • *Village in the City*, N. Taylor • *Essex Design Guide* • GLDP inquiry report • Liverpool offices public inquiry
1974	• Heath government collapses • Wilson PM		• London Docklands Joint Committee established • Covent Garden Forum established • Autonomous House model A. Pike

185

Date	General context	Building Industry developments	Architectural events
1975	• Referendum on Britain's entry to EEC • First oil from the North Sea	• New Architecture movement	• Secretary of State for Environment decision on GLDP • 'Camden Town Nine' letter to The Times re RCA • Sheffield begins housing improvement • Lambeth housing wins Heritage Year award • Secretary of State for Environment chairs Liverpool Partnership Committee • Willis Faber Dumas, N. Foster
1976	• Entebbe raid		• Manchester Royal Exchange Theatre, Levitt Bernstein • National Theatre, Lasdun
1977	• Conservatives win GLC election	• Publication of Inner Area Studies	• Language of Post-Modern Architecture, C. Jencks • RIBA allows advertising • Monopolies Commission Report • Highway engineering design rules changed • Pompidou Centre, Piano and Rogers
1978			• Morality in Architecture, D. Watkin • Covent Garden Plan • South Yorkshire Structure Plan • Hillingdon Civic Centre, RMJM • Alexandra Road, Camden
1979	• Conservatives win General Election	• Arts Council Thirties exhibition	• Birkenhead 'Piggeries' demolished • Philip Johnson, RIBA discourse • London Docklands Development Corporation set up • Pooley speaks on opportunity planning • Wimpey build private houses in Everton • Merseyside Draft Structure Plan • Newcastle Metro opened
1980	• John Lennon murdered	• Design and build contracts • Association of Consultant Architects (ACA) • Enterprise Zones established • Development. Corporations designated	• Parker-Morris standards abandoned • The Land Decade launched by A. Coleman • Milton Keynes shopping city
1981	• Toxteth riots • Heseltine proposes Merseyside Task Force	• RIBA code and ARCUK allow directorships	• Clifton Nurseries, T. Farrell
1982			• Lubetkin awarded RIBA Gold Medal • Inmos Factory, R. Rogers
1983		• RIBA agrees architects may advertise	• TVAM, T. Farrell • Burrell Collection, B. Gasson

186

Date	General context	Building Industry developments	Architectural events
1984	• Conservatives win General Election	• Prince of Wales Hampton Court speech • New Building Regulations proposed	• Liverpool Garden Festival, Arup • Classical Revival exhibition
1985		• EC Architects Directive • Mies design for Mansion House Square rejected	• National Gallery extension, new limited competition.
1986	• GLC abolished • Big Bang on Stock Exchange	• RIBA mandatory fee scale abolished • Plan to close architecture schools abandoned • Eurotunnel proposal launched	• Mansion House Square scheme by Stirling, Wilford • Plans for Canary Wharf announced • LLoyds', R. Rogers • Paternoster Square master plan competition.
1987	• Conservatives win General Election • Stock market crash		• Lords' Mound Stand, M. Hopkins • Rod Hackney President of RIBA
1988	• Housing Act		• Camden Town supermarket, N. Grimshaw • Proposal for King's Cross redevelopment, N. Foster • Canary Wharf redesign
1989		• Visions of Britain exhibition, Prince of Wales	• Richmond Riverside, Q. Terry
1990	• Enterprise Zone tax advantages expire	• Energy Conscious Design exhibition • ACA starts planning for Single European Market	• Imagination HQ, R. Herron

187

Appendix II

Methodology of the survey

In collaborating on developing the information for this volume, the authors recognized that they were on different sides of an ocean and, therefore, needed to arrange the work to accommodate that distance. (Although it is worth noting that with the increasing use of facsimile transmission and electronic mail linking universities, the effect of such distances is continually shrinking.) It was decided that the questionnaire survey would be developed primarily in the United States and the forms would be posted to architects in the United Kingdom to be returned to a UK address by the responding architects. They would then be forwarded to the US for entry into the computer and for analysis. After a review of the literature, a draft questionnaire was written during the spring of 1991. It travelled by facsimile between Texas and Manchester a number of times while it was under revision. There were a great many questions which begged to be asked. However, it was known that the questionnaire should be kept as short as possible in order to increase the potential response rate. The final questionnaire was only four pages in length but had 171 questions packed on to those few pages. We knew we were asking the responding principals to spend a significant period of time providing information.

In the autumn of 1991, a random sample of principals was taken from the Royal Institute of British Architects' directory of member firms.[1] It was a simple random sample of 1173 names. No more than one principal per firm was selected. Each member of the sample was mailed a questionnaire, a covering letter with the University of Manchester letterhead, and an envelope pre-addressed to the University of Manchester within which to return the questionnaire. As the completed questionnaires arrived in the mail, they were grouped and forwarded to Texas A&M University. A number of techniques were used to increase the response rate. While complete confidentiality was promised in the covering letter, with only summary statistics to be reported, the questionnaires were coded so we knew who had responded. Eight weeks after the initial mailing, a follow-up postcard was sent to those who had not yet responded requesting that they please answer. After another six weeks, a second letter, questionnaire and reply envelope was sent to those who had still not responded. In the end 610 responses were received. This represents a highly respectable response rate for a mailed questionnaire of 52 per cent. Of course, it must be acknowledged that a bias may have been introduced into the data by those who did not respond. As is always the case in research like this, the direction of that possible bias remains unknown. However, response rates on mailed surveys are rarely this high and, given that other available data had dramatically smaller samples and therefore even greater potential for unknown bias, and was substantially out of date, this rate of response was considered excellent. Of course, not every respondent answered every question. This is usual in surveys. In the data reported here, the percentages shown are of those responding to a particular item. The data were analysed using SPSS-PC+, an industry standard statistical analysis program. The draft chapters containing results were transmitted across the Atlantic, computer to computer, using the Internet system, and final editing was undertaken in Britain.

Before examining the survey's results, it is useful to define two terms: 'design' and 'style'. We wish there were simple definitions; unfortunately, there are not. These two terms, representing concepts central to the work of the architect, have become obscured and blurred. Their definitions and their usage have become highly overlapping.

Webster's Seventh New Collegiate Dictionary defines design in this context as 'to conceive and plan out... to have as a purpose... to devise for a specific function or end... to draw the plans for'.[2] It defines style as 'to design and make in accord with the prevailing mode'.[3] While the distinction can be seen in these definitions, the overlap is also evident. It has been made quite clear (most popularly by Tom Wolfe in his book *From Bauhaus to Our House*[4]) that the twentieth century modernist tradition was to consider style an inappropriate attribute for architecture. If buildings (and other objects) were properly designed, stylistic attributes would be unnecessary. 'Style' became a taboo word. Of course, what happened was that a visual style in fact evolved. Mies van der Rohe's well known and well worn exhortation, 'less is more', similarly evolved to a style. This is not necessarily bad. To these authors' minds, there is nothing wrong with an object, including a building, conforming to the visual, physical and formal rules of a style. However, the confusion occurs when we speak of that object being well 'designed'. Design, to these authors, conforming to the dictionary definition cited above, refers to the planning of a building. That planning includes the incorporation of knowledge from all areas appropriate to the task. Style should refer to the application of physical and visual attributes. Yet, the use of the word 'design' in the questionnaire was inescapable. It should be remembered by the reader that the word design may have had different meanings to different responding principals when each was completing the questionnaire.

Representativeness of the sample

Whenever a sample is used to collect data about a larger population, it is always useful to compare the responses received from the sample to whatever census data are available about the population in its entirety. In this case, the RIBA annually collects age and gender data about its members.

Table II.1 Testing the sample: comparison of population and sample age and gender data* (per cent)

| | Source and year | | | |
| | RIBA | RIBA | RIBA | This study |
Age: (%)	1990	1991	1992	1991
<30	1.0	1.5	1.5	0.8
30–34	8.7	6.5	6	5.4
35–39	14.9	14	15.5	11.5
40–44	20.1	20	18	19.2
45–49	16.5	17	20	20.8
50–54	12.6	13	16	12.6
55–59	12.9	12.5	10.5	12.3
60–64	10.0	13	9	10.5
65+	3.3	3.5	3.5	6.9
Gender:				
Male		96.8	96.0	95.9
Female		3.2	4.0	4.1

*   The 'RIBA data' listed above had to be calculated from the data reported by the RIBA. For example, the RIBA data for 1991 and 1992 listed sole principals and principals in partnerships separately. For simplicity of comparison, they are combined here. The gender distribution data distributed by the RIBA showed the percentage of females who are principals and the total percentage of architects who are female. The percentage of principals who are female, the figure needed here for accurate comparison with the sample, was calculated from those RIBA data. It was not possible to calculate the percentage of principals who are female from the RIBA report for 1990.

If the RIBA's census data closely resemble the data obtained by the sample, it is a good indication that the sample fairly represents the larger population. Table II.1 shows the RIBA's age data for 1990, 1991 and 1992, as well as the distribution obtained from the sample used here.[5] Regarding the age data from the RIBA, an interesting fact immediately appears. The 1990, 1991 and 1992 age distributions all very closely match the age distribution of the sample used for this study. To these researchers this lends substantial credibility to the opinion that the results reported here are representative of principals at large. The gender data from the RIBA clearly also confirm the representative nature of the sample.

Demographics of the responding principals

The sample for this study was drawn from a directory of RIBA member firms, as indicated earlier. All of these firms are in private sector practice. Given that sampling universe, the range of ages of the responding principals is what one would expect. The largest group of principals were in their 40s followed by the group in their 50s. A few firms with relatively young (under 30) principals were found by the sample, as one would expect. Also, a few principals beyond the usual retirement age, as one would expect, were also found.

Table II.2 Age of respondents

	Group	No.	%
29 and under	1	5	0.8
30–39	2	103	16.9
40–49	3	244	40.0
50–59	4	152	24.9
60–69	5	94	15.4
70 and over	6	12	2.0
Total:		610	100.0
Mean=44.3 years (3.431)			
Valid cases 610	Missing cases 0		
Regrouped for analysis:			
Up to 50		365	59.8
51–65		209	34.3
66+		36	5.9

However, the question immediately arises of whether differences in age are associated with differing views of the profession. For this analysis, responding principals were regrouped into three categories: those 50 years old or younger, those 51 to 65 years old and those over 65, a common retirement age. From examining the data, age 50 appears to be the point most likely to show differences between those older and younger. These differences have been examined elsewhere in this report.

Gender

Nearly all of the responding principals were male, as shown in Table II.3. One wishes that there were more women in the profession. It has only been over the last few decades that women have entered the profession in large numbers. One can, of course, argue that sufficient time has not yet passed for these women to attain the principal's position.

Table II.3 Gender of respondents

	Group	No.	%
Male	1	582	95.9
Female	2	25	4.1
Total		610	100.0
Valid cases 607, missing cases 3			

None the less, for under one-twentieth of the principals to be women, given the high percentage of women now in architecture university programmes, shows a profession about to be hit with a rapid change in this area. Also, one wishes that an RIBA census of women principals were available. This would provide an interesting check on the validity of the sample. Lastly, did the women have different views of the profession than the men? Unfortunately, the sample size was too small for meaningful statistics to emerge.

Distribution by firm size

Firm size has become a more difficult question over recent decades. First, one must remember that the sample was drawn from a roster of RIBA member firms. Therefore, public bodies were excluded. Also, a very large number of the architects in the sample work in firms with very varied professional staff members. They work alongside engineers, urban planners, landscape architects, interior designers and other professionals. But, how do we measure the size of a firm in this context? Is it the number of architects? Is it the number of design-related professionals? Is it the number of all professionals? Is it the total level of billings? Given that the architect is increasingly working as a member of a team of varied professionals, the question of the role of firm size becomes more and more perplexing. The RIBA 1962 study examined firm size in two ways.[6] It used a three-way split (10 or fewer architects, 11 to 30 architects, and over 30 architects) and a five-way split (5 or fewer architects, 6 to 10, 11 to 30, 31 to 50 and over 50 architects in the firm). We report each of these separately in Tables II.4 and II.5.

What do these data tell us? On first glance, these two tables would indicate that architects tend to work in very small firms and do not have joint practices with other design-related professions. However, this may not be the case.

To investigate this further, we created two additional variables. The first we called the multi-disciplinary practice. Excluding the technicians, since they do not necessarily represent a different qualified profession, practices for which responding principals reported that their firm had at least one architect and at least one other member from another design or construction profession were categorized as multi-disciplinary practices.

Table II.4 Number of qualified professionals in firms: five-way split (no. (per cent))

	None	1–5	6–10	11–30	31–50	51+
Architects	34 (5.6)	451 (73.9)	60 (9.8)	50 (8.2)	11 (1.8)	4 (0.7)
Interior designers	474 (77.7)	129 (21.1)	5 (0.8)	2 (0.3)		
Planners	523 (85.7)	83 (13.6)	4 (0.7)			
Quantity surveyors	566 (92.8)	39 (6.4)	5 (0.8)			
Engineers	573 (93.9)	29 (4.8)	2 (0.3)	3 (0.5)	3 (0.5)	
Technicians	208 (34.1)	318 (52.1)	39 (6.4)	32 (5.2)	8 (1.3)	5 (0.8)
Landscape architects	558 (91.5)	47 (7.7)	5 (0.8)			
Other design disciplines	557 (91.3)	50 (8.2)	3 (0.5)			

Table II.5 Number of qualified professionals in firms: three-way split (no. (per cent))

	None	1-10	11-30	31+
Architects	34 (5.6)	511 (83.8)	50 (8.2)	15 (2.5)
Interior designers	474 (77.7)	134 (22.0)	2 (0.3)	
Planners	523 (85.7)	87 (14.3)		
Quantity surveyors	566 (92.8)	39 (6.4)	5 (0.8)	
Engineers	573 (93.9)	31 (5.1)	3 (0.5)	3 (0.5)
Technicians	208 (34.1)	357 (58.5)	32 (5.2)	13 (2.1)
Landscape architects	558 (91.5)	52 (8.5)		
Other design disciplines	557 (91.3)	53 (8.7)		

From Table II.6 we can see that there are an impressive number of multi-disciplinary practices: over 35 per cent of the firms are multi-disciplinary.

Table II.6 Multi-disciplinary practices:
firms having at least one architect and
at least one other design professional
(per cent (no.))

No	64.6 (394)
Yes	35.4 (216)

The second variable we created was to examine firm size not only by the number of architects but also by the number of design-related professionals. In Table II.7 we grouped the counts of all the different professionals together. This seemed quite reasonable if we assume that the reason we are interested in the size of a firm is that larger firms permit greater specialization. Whether or not all professionals are architects is not necessarily the only item of interest. This told us something rather different about firm size.

Table II.7 Firm size including all design professions

Five-way split:	Frequency	Percent
1–5	416	71.5
6–10	85	14.6
11–30	58	10.0
31–50	8	1.4
51+	15	2.6
Three-way split	**Frequency**	**Percent**
1–10	501	86.1
11–30	58	10.0
31+	23	4.0
Valid cases 582	Missing cases 28	

None the less, even when the question of firm size is examined in this multidisciplinary way and even when given that over one-third of RIBA member firms are multi-disciplinary practices, one still sees a picture of very small firms. Of those responding, 24.6 per cent are in firms with only one professional. An additional 21.8 per cent are in firms with only two. Another 12.7 per cent are in firms with only three design professionals. This totals to the fact that 59.1 per cent of those responding are in firms with three or fewer professionals. Architecture may be one of the last cottage industries. It should be noted that two responding principals reported that their firms had 131 professionals in total, and 2.1 per cent of those responding indicated that their firms had 61 or more professionals from all design disciplines. So, not every firm is small. A sizeable number of practices are multi-disciplinary. But, the preponderance of firms are tiny and represent only a single discipline of the many areas of expertise needed in the building process.

There is another, and final, possible way of measuring firm size, as shown in Table II.8. It is the total value of buildings produced on average each year. The responding principals were asked, 'Approximately how much business does your firm do per year in terms of the value of buildings?' The questionnaire provided a blank line for the amount of pounds sterling to be filled in. While one responding principal wrote, 'varies wildly', the responding principals overall filled in a number. The economy had moved slowly into a recessionary period, as it may now be extremely slowly moving out of it. Yet, the year prior to the survey's distribution, which these numbers are most likely to represent, was probably neither a high nor a low. Again, however, the point is not whether this number accurately represents an average. It is whether it is an accurate surrogate measure for firm size.

Table II.8 Firm size measured by size of annual billings

Total billings	Frequency	%
<£0.5 million	113	18.5
£0.5–£1 million	53	8.7
£1–£8 million	295	48.4
> £8 million	149	24.4
Mean=£0.939 million		
Valid cases 610		

This figure, at least on the surface, does not seem to accurately represent firm size. While the preponderance of firms were very small with regard to either the number of architects or the total number of professionals, nearly three-quarters of these firms completed buildings greater than £1 million. Now, of course, one must remember that £1 million does not necessarily purchase that much of a building in today's world. But, the fees or commissions earned on such a structure would provide a reasonable income for at least a firm with two architects. Yet, the value of buildings produced does not seem to be a good surrogate for firm size.

Elsewhere we examine whether firm size appears to make a difference in how those responding view other factors in the practice of architecture.

Comparisons of this study with the 1962 RIBA study

Comparisons with the 1962 study are exceptionally problematic. The RIBA study used a smaller sample which totals 258. Some offices were visited, others were sent a mail questionnaire. It is not clear to the authors whether the participating offices were chosen randomly. Some were chosen not for representativeness but because they were thought to indicate excellence. While this might be useful for purposes of emulation, it does not give a picture of the entire profession. Furthermore, the data were not reported consistently. The 1962 study gives the number of offices with 11 and more architects who received the mailed questionnaire. And, when discussing offices visited, the number of offices is broken down into 11 to 30, 31 to 50 and over 50 architects in the office. It is unfortunate that more of the statistical data was not published at this level of detail.

One must also note that the 1991 data uncovered 34 principals of RIBA member firms who reported that there were no qualified architects within their firms. There are a number of factors which could explain this. The most likely is that the directory that was used to draw the sample was a year old. Some firms that appeared in the sample may have dropped the services of an architect as a part of their portfolio of services. But, any correctly administered random sampling procedure would identify a representative number of these firms. Still, acknowledging that, one can compare the sample sizes.[7]

Table II.9 Comparing the 1962 and 1991 samples

	No. of architects														
	0		1–5		6–10		11–30		31–50		50+		11+		Total	
	1962	1991	1962	1991	1962	1991	1962	1991	1962	1991	1962	1991	1962	1991	1962	1991
Visited offices			6		9		14		7		8		34		78	
Mail questionnaire	34		138	451	42	60		50		11		4			180	610
Total	34		144	451	51	60	14	50	7	11	8	4	34		258	610
% of total	7		56	74	20	10	5	8	3	2	3	1	13		100	100

From Table II.9 one sees, over the nearly 30-year period, a nearly 50 per cent growth in the proportion of small offices (1 to 5 architects); a 50 per cent reduction in offices with 6 to 10 architects; and, generally, an over 50 per cent reduction in the number of offices with 11 or more architects (1962, 24 per cent; 1992, 11 per cent). If the numbers can be considered representative, we see a substantial growth in the number of small firms. In a 1988[8] survey 69.2 per cent of practices have 1 to 5 architects, 16.4 per cent have 6 to 10 architects, 10.9 per cent have 11 to 30 architects, 1.9 per cent have 31 to 50 architects, and 1.5 per cent have over 51 architects. These figures (which were not from a sample but from the full RIBA membership) generally confirm the numbers reported here for the 1991 sample. This lends further credence to the representativeness of the 1991 sample generally.

Principals have a longer-range view

One might ask, 'Why did they send the questionnaire only to private sector principals?' This, of course, is a reasonable question. Yet, there are clear reasons why this route was chosen. Architects work either for the private sector or for the public sector. While both sectors are facing challenges, the issues confronting an architect in the public sector may be very different from those facing the architect in the private sector. For example, it is well known that public agencies may be increasingly utilizing the architect on staff in a liaison or supervisory role with the private sector architect in his or her own firm who is actually designing the building. These are very different tasks.

197

The architect employed in the public sector would, again for example, be much more likely to be spending time on management-related activities rather than design-related activities. It was clear that in order to obtain statistically significant results from the data, a large sample would be needed. But, if both public sector and private sector architects were included, this would effectively reduce the size of each group, given finite resources. The decision was made to concentrate only on architects employed in the private sector.

Additionally, only principals received the survey. Principals are most likely to have a longer-range view of the profession and more likely to be focusing on the trends for the future of the profession than salaried employees. Since the questionnaire was only sent to principals and such a high response rate was achieved, it can be said that the results extremely closely follow what the results would have been had the entire universe of principals of architecture firms been questioned. Such are the benefits of statistical analysis. You can question a smaller group (the sample) and learn, within 2 or 3 per cent, what the views are of the larger group (the universe). Also, it is quite likely that a principal in a firm has a different view from an entry-level assistant. Yet, the principal has had a number of years of experience, is sufficiently distant from his or her university education to make independent judgements, and has gained perspective on the profession and its pattern of changes. Principals could provide more wisdom as well as a more historical and longer-range view than juniors.

The questionnaire itself is shown in Appendix III. As the reader can see, many questions were about the principal himself or herself and many questions were about his or her firm. Yet, there was very little ambiguity in each of these questions. The authors believe that it is very clear, from the nature of the question, whether the respondent is answering about the firm or the individual. Most questions, of course, queried the principal on his or her view. The researchers have no way of knowing if the responding principal passed the questionnaire to a junior associate for a response, of course. The assumption must be made that the principal, himself or herself, answered the questionnaire.

Lastly, using the directory of RIBA member firms obviously excluded principals who are employed with firms which are not RIBA members. In effect, this means that the results reported only the views of principals in RIBA member firms. However, by narrowing the focus to this group the results have gained statistical accuracy. The results represent a longer-range view and, it is hoped, a wisdom about the profession. It must be left to another study to examine architects working in the public sector, the views of less senior architects and of those not members of the RIBA.

Interpreting data tables of the differences sections

Any presentation of questionnaire survey results necessitates the inclusion of numbers. Every effort has been made to minimize the presentation of raw numbers in Part Two, yet, of course, numbers still abound. The tables presented in the various sections discussing 'differences' (differences by firm size, differences by whether a practice is multi-disciplinary, etc.) are only summaries, not the full results. It was felt that some readers might benefit from a few words of explanation on how to interpret these tables. There is actually a good deal more information embedded in each table than meets the eye on first glance. As an example, take Table 5.17, which looks at the time spent by principals on certain activities, differentiated by whether their practices are multi-disciplinary. The first line of the table shows that:

1. of those principals who are not in multi-disciplinary practices, 36.3 per cent (143 responses) spend 20 per cent or more of their time on building design;

2. of those who *are* in multi-disciplinary practices, 32.9 per cent (71 responses) spend 20 per cent or more of their time on building design.

The table also gives the reader important information by what it does not show but is embedded in the information given. For example, it indicates that of those principals who are not in multi-disciplinary practices, 63.7 per cent spend less than 20 per cent of their time on building design; and of those principals who are in multi-disciplinary practices, 67.1 per cent spend less than 20 per cent of their time on building design. This would lead to the conclusion that principals in multi-disciplinary practices are less likely to spend time on building design. The authors, of course, faced choices in how to present the data. All of the data could have been presented. However, this would have yielded tables that are about four times as long. The summary tables were chosen because they present the most important information without giving the reader 'table overload'. If that overload is still present, please accept our apologies. The tables that remain were demanded by the usual standards for the thorough presentation of research results. It should also be noted that within each 'differences' section, *all* possible differences were run on the computer. Unless otherwise noted, only those which showed statistically significant differences at $p < 0.05$ are included in this report.

The chi-squared test was used. This indicates that there is less than a 5 per cent likelihood that these results would occur randomly. It can be assumed by the reader that if a difference is not reported, it did not show as statistically significant in the computer analysis.

Notes

1. *RIBA Practices '89: A Directory of RIBA Practices*, London, RIBA, 1989.

2. *Webster's Seventh New Collegiate Dictionary*, Springfield, MA, G&C Merriam, 1965, p. 224.

3. Ibid., p 872.

4. Wolfe, T. *From Bauhaus to Our House*, New York, Farrar Straus Giroux, 1981

5. *Architects' Employment and Earnings*, London, RIBA, 1990, 1991, 1992.

6. *The Architect and His Office*, London, RIBA, 1962.

7. Ibid, p 28.

8. RIBA Market Research Unit, *Census of Practices*, RIBA, 1989.

Appendix III

Study of the profession questionnaire

STUDY OF THE PROFESSION QUESTIONNAIRE

We are conducting a survey of the activities of architects in the United Kingdom. Simply circle the number indicating your response to each item. Your answers will be confidential and only summary results will be published. Upon completion of each questionnaire, please return it in the envelope provided. Thank you for your help.

PART I.
What approximate percentage of the time over the past year has your firm been employed on the following things:

[1] UP TO 10% [2] FROM 11-25% [3] FROM 26-50% [4] MORE THAN 50%

1.Housing projects .... [1] [2] [3] [4]   4.Urban design.............. [1] [2] [3] [4]
2.Commercial and industrial building .. [1] [2] [3] [4]   5.Individual clients........ [1] [2] [3] [4]
3.Institutional and public buildings ..... [1] [2] [3] [4]   6.Other (please specify)_____

PART II.
We are interested in the time you spend in a typical week for a variety of activities. Please estimate as closely as possible the amount of time per week you spent on the following tasks, according to the scale as shown. Circle the appropriate number to indicate your answer.

	<½ DAY	½ DAY	1 DAY	1½ DAYS	2 DAYS	2½ DAYS	3 DAYS	4 DAYS	5or> DAYS
MANAGING CURRENT PROJECTS:									
7. Coordinating consultants and in-house staff	[1]	[2]	[3]	[4]	[5]	[6]	[7]	[8]	[9]
8. Meeting with clients	[1]	[2]	[3]	[4]	[5]	[6]	[7]	[8]	[9]
9. Meeting with project managers	[1]	[2]	[3]	[4]	[5]	[6]	[7]	[8]	[9]
10. Getting agreements in writing	[1]	[2]	[3]	[4]	[5]	[6]	[7]	[8]	[9]
11. Writing accurate specifications	[1]	[2]	[3]	[4]	[5]	[6]	[7]	[8]	[9]
12. Construction site supervision	[1]	[2]	[3]	[4]	[5]	[6]	[7]	[8]	[9]
13. Monitoring construction budgets	[1]	[2]	[3]	[4]	[5]	[6]	[7]	[8]	[9]
14. Building design	[1]	[2]	[3]	[4]	[5]	[6]	[7]	[8]	[9]
15. Building production drawings	[1]	[2]	[3]	[4]	[5]	[6]	[7]	[8]	[9]
16. Estimating amount of work remaining prior to completion of a project	[1]	[2]	[3]	[4]	[5]	[6]	[7]	[8]	[9]
17. Other activities (please specify)_____									
GETTING NEW WORK:									
18. Publicizing completed work	[1]	[2]	[3]	[4]	[5]	[6]	[7]	[8]	[9]
19. Working with marketing specialists or public relations consultants	[1]	[2]	[3]	[4]	[5]	[6]	[7]	[8]	[9]
20. Recruiting clients by phone	[1]	[2]	[3]	[4]	[5]	[6]	[7]	[8]	[9]
21. Recruiting clients by mail	[1]	[2]	[3]	[4]	[5]	[6]	[7]	[8]	[9]
22. Recruiting clients in person	[1]	[2]	[3]	[4]	[5]	[6]	[7]	[8]	[9]
23. Negotiating/contracting new work	[1]	[2]	[3]	[4]	[5]	[6]	[7]	[8]	[9]
24. Other activities (please specify)_____									
RUNNING THE OFFICE:									
25. Staffing the office/keeping the right employees	[1]	[2]	[3]	[4]	[5]	[6]	[7]	[8]	[9]
26. Developing office procedures	[1]	[2]	[3]	[4]	[5]	[6]	[7]	[8]	[9]
27. Establishing the office fee structure	[1]	[2]	[3]	[4]	[5]	[6]	[7]	[8]	[9]
28. Managing office finances	[1]	[2]	[3]	[4]	[5]	[6]	[7]	[8]	[9]
29. Other activities (please specify)_____									

PART III.
The following statements will ask for information about meetings in your practice. Please respond according to the scale shown. Circle the appropriate letter to indicate your answer according to the scale:

[1] I STRONGLY DISAGREE   [2] I DISAGREE   [3] I DON'T KNOW   [4] I AGREE   [5] I STRONGLY AGREE

30. The preliminary meetings between the client and the architect are the most crucial ones .... [1][2][3][4][5]
31. Informal, unscheduled meetings with colleagues are more rewarding than formal, scheduled meetings .... [1][2][3][4][5]

[1] I STRONGLY DISAGREE   [2] I DISAGREE   [3] I DON'T KNOW   [4] I AGREE   [5] I STRONGLY AGREE

32. Numerous meetings should take place between architects and clients
during all phases of a project .... [1][2][3][4][5]
33. The architect should meet separately with each participant (client,
engineer, manager, etc.) in a project .... [1][2][3][4][5]
34. Meetings should be used to get client approval only after design issues
have been decided by the architect .... [1][2][3][4][5]

How often do your meetings take place at these locations:
[1] UP TO 10% OF MEETINGS   [2] FROM 11-25% [3] FROM 26-50%   [4] MORE THAN 50%

35. Over lunch   [1][2][3][4]   38. At the project site   [1][2][3][4]
36. On the phone   [1][2][3][4]   39. In the client's office [1][2][3][4]
37. In your office   [1][2][3][4]

PART IV.

Please respond to the following by indicating how important each item is to your work as an architect. Circle the appropriate number to indicate your answer, according to the following scale:

[1] VERY UNIMPORTANT [2] UNIMPORTANT [3] I DON'T KNOW [4] IMPORTANT [5] VERY IMPORTANT

HOW IMPORTANT IS:

40. Joining client trade associations or business and service organizations to
establish rapport with clients .... [1][2][3][4][5]
41. Making clear in publicity material that you will work in a subsidiary design
role if necessary .... [1][2][3][4][5]
42. Talking regularly with land surveyors, engineers, contractors and other to
get leads on clients .... [1][2][3][4][5]
43. Keeping up with real estate news and local planning reports to learn about
possible new clients .... [1][2][3][4][5]
44. Visiting and leaving information with potential clients to establish a
rapport before they need an architect .... [1][2][3][4][5]
45. Calling clients informally about some item outside of architecture of
architecture of interest to the client .... [1][2][3][4][5]
46. Inviting the client out socially .... [1][2][3][4][5]
47. Direct contact between the user and the architect during the design
process .... [1][2][3][4][5]
48. Hiring new staff because they have a similar philosophy of architecture to
others in your office .... [1][2][3][4][5]
49. Hiring new staff based on their talent as designers .... [1][2][3][4][5]
50. Hiring new staff with a wide variety of capabilities .... [1][2][3][4][5]
51. Globalization of the profession .... [1][2][3][4][5]
52. Law and liability .... [1][2][3][4][5]
53. Professional codes of ethics .... [1][2][3][4][5]
54. Cultural diversity of clients .... [1][2][3][4][5]
55. Technological innovation .... [1][2][3][4][5]
56. Land usage in design .... [1][2][3][4][5]
57. Application of research to practice .... [1][2][3][4][5]
58. Continuing education .... [1][2][3][4][5]
59. Habitability of the completed structure .... [1][2][3][4][5]
60. Client satisfaction .... [1][2][3][4][5]
61. Visual aesthetics of the building .... [1][2][3][4][5]
62. Education of the client and the public .... [1][2][3][4][5]
63. Promotion of architectural thinking .... [1][2][3][4][5]
64. Project management .... [1][2][3][4][5]
65. Construction management .... [1][2][3][4][5]
66. Avoidance of conflict with the client, owner or contractor .... [1][2][3][4][5]
67. Engineering .... [1][2][3][4][5]
68. Serving people's needs .... [1][2][3][4][5]
69. The chance to be creative .... [1][2][3][4][5]
70. Public recognition .... [1][2][3][4][5]

PART V.
This section asks questions about what architecture schools teach. On the left side we ask about what you think architects should learn, and on the right side we ask about what you actually studied in school. Please circle the appropriate number to indicate your answer according to the following scale:

[1] I STRONGLY DISAGREE [2] I DISAGREE [3] I DON'T KNOW [4] I AGREE [5] I STRONGLY AGREE

AN ARCHITECT NEEDS TO RECEIVE TRAINING IN:			I RECEIVED ADEQUATE TRAINING IN:	
71.	[1] [2] [3] [4] [5]	office management	95.	[1] [2] [3] [4] [5]
72.	[1] [2] [3] [4] [5]	marketing	96.	[1] [2] [3] [4] [5]
73.	[1] [2] [3] [4] [5]	accounting	97.	[1] [2] [3] [4] [5]
74.	[1] [2] [3] [4] [5]	schematic design	98.	[1] [2] [3] [4] [5]
75.	[1] [2] [3] [4] [5]	budget management	99.	[1] [2] [3] [4] [5]
76.	[1] [2] [3] [4] [5]	human behaviour	100.	[1] [2] [3] [4] [5]
77.	[1] [2] [3] [4] [5]	history of architecture	101.	[1] [2] [3] [4] [5]
78.	[1] [2] [3] [4] [5]	computer-aided design	102.	[1] [2] [3] [4] [5]
79.	[1] [2] [3] [4] [5]	building technology	103.	[1] [2] [3] [4] [5]
80.	[1] [2] [3] [4] [5]	client relations	104.	[1] [2] [3] [4] [5]
81.	[1] [2] [3] [4] [5]	communication	105.	[1] [2] [3] [4] [5]
82.	[1] [2] [3] [4] [5]	urban design/planning	106.	[1] [2] [3] [4] [5]
83.	[1] [2] [3] [4] [5]	construction management	107.	[1] [2] [3] [4] [5]
84.	[1] [2] [3] [4] [5]	project management	108.	[1] [2] [3] [4] [5]
85.	[1] [2] [3] [4] [5]	facility management	109.	[1] [2] [3] [4] [5]
86.	[1] [2] [3] [4] [5]	interior design	110.	[1] [2] [3] [4] [5]
87.	[1] [2] [3] [4] [5]	real estate development	111.	[1] [2] [3] [4] [5]
88.	[1] [2] [3] [4] [5]	structural/mechanical design	112.	[1] [2] [3] [4] [5]
89.	[1] [2] [3] [4] [5]	specifications and codes	113.	[1] [2] [3] [4] [5]
90.	[1] [2] [3] [4] [5]	brief preparation	114.	[1] [2] [3] [4] [5]
91.	[1] [2] [3] [4] [5]	production	115.	[1] [2] [3] [4] [5]
92.	[1] [2] [3] [4] [5]	research	116.	[1] [2] [3] [4] [5]
93.	[1] [2] [3] [4] [5]	computerization	117.	[1] [2] [3] [4] [5]
94.	[1] [2] [3] [4] [5]	other (please specify) _____	118.	[1] [2] [3] [4] [5]

PART VI.
The following are statements about your experience in practice, and your beliefs about certain issues. Please circle the appropriate number to indicate your answer, using the following scale:

[1] I STRONGLY DISAGREE [2] I DISAGREE [3] I DON'T KNOW [4] I AGREE [5] I STRONGLY AGREE

119. The responsibility falls to the architect to see that the contractor carries out the plans as promised to the client .... [1] [2] [3] [4] [5]
120. The desire to produce excellent projects is often overshadowed by clients limitations and demands .... [1] [2] [3] [4] [5]
121. Architects always place public interest above client satisfaction .... [1] [2] [3] [4] [5]
122. Private connections and contacts often lead our firm to clients .... [1] [2] [3] [4] [5]
123. I frequently work in the evenings and on weekends .... [1] [2] [3] [4] [5]
124. It is difficult to find more than 30 minutes of uninterrupted time in a working day.. [1] [2] [3] [4] [5]
125. Setting a workable time schedule for projects is a top priority for my firm [1] [2] [3] [4] [5]
126. Most architects adhere to strict professional codes of ethics .... [1] [2] [3] [4] [5]
127. Architects should always be prepared to offer only a partial service (design only) and allow others to lead the team .... [1] [2] [3] [4] [5]
128. Only the owner or principal of the firm should determine fees................. [1] [2] [3] [4] [5]
129. Only the owner or principal of the firm should evaluate the work of associates... [1] [2] [3] [4] [5]
130. Negotiations with clients should be handled only by the owner or principal of the firm .... [1] [2] [3] [4] [5]
131. It is not always possible to know the tasks a project will involve before beginning work on it... [1] [2] [3] [4] [5]
132. Radical or innovative ideas are interesting and thought-provoking .... [1] [2] [3] [4] [5]
133. Such ideas are, in the long run, best avoided in favour of the more conventional.. [1] [2] [3] [4] [5]
134. Personal style is important to the success of an architect........................ [1] [2] [3] [4] [5]
135. Architects should uphold their design standards, regardless of the fee involved.. [1] [2] [3] [4] [5]

205

[1] I STRONGLY DISAGREE   [2] I DISAGREE   [3] I DON'T KNOW   [4] I AGREE   [5] I STRONGLY AGREE

136. The architect and the client should reach agreement about the project
early in schematic design... [1] [2] [3] [4] [5]
137. In writing contracts, some arrangements should be left open-ended in
hopes that the job will grow .... [1] [2] [3] [4] [5]
138. Efficiency is more important than creativity in designing a project........... [1] [2] [3] [4] [5]
139. Architecture today is more a business enterprise than a profession......... [1] [2] [3] [4] [5]
140. Firms which focus entirely on design have become obsolete .... [1] [2] [3] [4] [5]
141. The increased demand for architectural services over the last decade gave
me greater choice in projects... [1] [2] [3] [4] [5]
142. Greater competition in the architectural field has forced me to explore new
approaches to architectural practice .... [1] [2] [3] [4] [5]
143. Computers will be vital to architecture in the future.............................. [1] [2] [3] [4] [5]
144. Today's architectural firms must offer more comprehensive services than
were necessary in the past.. [1] [2] [3] [4] [5]
145. If starting over, I would become an architect again .... [1] [2] [3] [4] [5]
146. I would be willing to work outside of architecture for more money .... [1] [2] [3] [4] [5]
147. I could foresee myself moving completely out of the architecture field for
my livelihood, for reasons other than illness or retirement...................... [1] [2] [3] [4] [5]
148. I am very satisfied with my career .... [1] [2] [3] [4] [5]

PART VII.
In this section we ask for some background information to help us analyse the results of this survey. Please circle the number indicating your response or fill in the blank space provided with your answer.

How many employees in your firm are qualified as:
149. Architects_____   153. Engineers_____
150. Interior Designers_____   154. Technicians_____
151. Planners_____   155. Landscape Architects_____
152. Quantity Surveyors   156. Other design disciplines_____
157. How may administrative staff members (ie secretarial and financial) are employed by your
firm?_____
158. Approximately what percentage of your business was as sub-contractor (eg. design subcontractor) 1989/90?_____
159. What geographic region is handled by your firm?
[1] City-wide [2] Regional [3] National [4] International
160. What is the primary focus of your firm's projects?
[1] Corporate [2] Public sector [3] Small-scale private [4] Mixed
161. Approximately how much business does your firm do per year in terms of the value of the
buildings? £_____
162. In the financial year 1989/1990, your office was: [1] very busy. [2] fairly busy. [3] not busy.
163. How many years have you been a registered architect? _____years.
164. By how many architectural firms have you been employed in your career, including your
present firm?_____ firms.
165. How long has your present firm been in existence?_____ years.
166. What is your sex? [1] Male [2] Female
167. What is your age?_____ years.
168. What college or university did you attend? _____
169. What professional qualifications do you hold? _____
170. What year did you graduate? _____
171. Is there anything else you would like to add about the skills you need in your practice, or
the skills you feel will be needed in the future? (Attach additional pages if necessary.)

Please return this questionnaire in the enclosed pre-addressed envelope. If the envelope is missing, please return your questionnaire to:

Andrew Seidel, Professor
c/o Martin Symes, Professor
University of Manchester
School of Architecture
Manchester, M13 PL9, United Kingdom

Appendix IV

Questionnaires used in semi-structured interviews

Interview with principals

- Company size now, address, phone.
- Interviewee, respondent, date.
- Fundamental questions: what do you do, how do you go about it, and why?

1 Origins

- Who started the practice?
- What training background, qualifications, registration?
- What connections with co-principals or other firms before?
- When/how did the practice start? Around any particular event/project? For example:
  - buying
  - inheriting
  - working up to take over a practice
  - one partner starting, or all joining together
  - competition win.
- Location:
  - Where did it start operating?
  - How many locations now, where?
  - Image requirements of premises?

2 Current professional structure

- Partnership, private, public company, other?
- How many employees now?
- How many are qualified architects, technicians, students, planners, engineers, administrative, other?
- Hierarchy and career structure in office?
- Age structure?
- Level mix?
- Other related companies started or associated? When, why?

3 Development and growth

- What where the above characteristics like at different points in the period of the study: 1968–1990? For example:
  - rate of growth/decline
  - staff turnover patterns

        change in mix of staff types.
- What events in or beyond the firm were benchmark moments? For example:
  - big projects
  - different type of work
  - recession
  - change among principals
  - competition
  - (Develop these and their reasons if possible.)

4 Projects

- Any describable shift over time?
- What skills/services are offered to clients?
- What projects are done? For example:
  - housing (public, housing association, private)
  - offices
  - industrial
  - retail
  - hotels
  - sport/leisure
  - health
  - religious
  - educational
  - urban design
  - product design
  - space planning
  - interiors
  - landscape
  - other consultancy.
- Large, medium, small (annual fee income)?
- Number running at once?
- New build, rehabilitation?
- Client types?
- Reasons for seeking or rejecting particular client/project types?

5 Management/organization type?

- Conscious management structure?
- If so, developed when, why, by what process?
- Type and frequency of meetings of which people?
- Centralized type: one (or more) groups with design control by principal(s)

little delegation of responsibility or job continuity?
- Or dispersed type: groups (static or fluid) each handling a set of projects?
- Is there an office manager?
- Architect, financial, management trained?
- Are decisions individualist or collective?
- Who makes policy decisions? For example, who decides:
>which jobs to look for/and take on
>how to court clients
>how budget control methods are handled
>whether and how to seek foreign work
>whether to team up with other professionals
>staff salary structures/and career paths?
- Recruitment methods: e.g. advertising? Other reasons for employee choice: e.g. wide skills, design flair, compatible design approach?
- Type of training offered to staff?
- Decisions on these matters taken when, why?

6 Working methods

- What services and skills are offered (see question 2)?
- What are the reasons for providing these services?
- What, if any, control systems are in place in project design and subsequent stages: e.g. site, cost control?
- What information systems are used: e.g. library, Barbour Index?
- Systematic in-house dissemination of experience etc?
- What technical systems are used: CAD, standardized details, specifications etc?
- What technical expertise exists in-house: e.g. engineering, cost?
- How is specialist advice on technical matters obtained:
>in-house
>nominated sub-contractors or suppliers
>other consultants paid by client or architect?
- Publicity and marketing methods:
>article writing
>cold calls
>other client contacts
>competition entries?
- What is the chosen knowledge base?
- What has to come from outside?
- When have new elements been added to the chosen base?

7 Philosophy (seek expansion of any of the above information)

- Reasons behind:
  > design methods/approaches
  > type of staff sought/and training given
  > type of clients sought/and rejected
  > relationships developed within and outside the firm
  > chosen knowledge base?

Information about projects

1 Project description

Name of project; building type; what was produced.
Client.
Dates start, finish.
Cost.
Pattern of progress.

2 Team

Who was team leader? Relation to other principals.
Size of team at different points in the project.
Make-up of team: special skills required.
Management structure of team.

3 Special characteristics in relation to firm's development

Challenges presented and how they were tackled.
Special technical skills required

4 Client relations

How the client came to the firm.
Type and frequency of interaction.
Degree of satisfaction.

5 Design development

Who led?
How was it refined within the team?
Part played by other members of the office.

6 Working methods

See question 6 in schedule for interview with principals: all questions as applied
to this project.

Bibliography

Ambasz, E. (1988) *Emilio Ambasz: The Poetics of the Pragmatic*, New York, Rizzoli.

American Institute of Architects *Personnel Practices Handbook (2nd edn.)*, Washington, DC, AIA.

American Institute of Architects *Architect's Handbook of Professional Practice.*

American Institute of Architects *The Federal Marketplace: Are You Prepared?*, Washington, DC, AIA.

American Institute of Architects (1970) *Financial Management for Architectural Firms: a Manual of Accounting Procedures*, Washington, DC, AIA.

American Institute of Architects (1971) *Financial Management for Architectural Firms: a Manual for Computer Users*, Washington, DC, AIA.

American Institute of Architects (1976) *Current Techniques in Architectural Practice*, Washington, DC, AIA.

American Institute of Architects (1978) *Standardized Accounting for Architects*, Washington, DC, AIA.

Aravot, I. and Seidel, A. D. (1994) The lost 'Architects' Desk Reference' and the need for rewriting it. In *Magazine of the Faculty of Architecture and Town Planning*, Haifa, Institute of Technology, pp. 22–31.

Arts and Architecture Press (1986) *Design Office Management Handbook*, Santa Monica, CA, AAP.

Auger, B. (1972) *The Architect and The Computer*, New York, Praeger.

Bachner, J. P. (1991) *Practice Management for Design Professionals: a Practical Guide to Avoiding Liability and Enhancing Profitability*, New York, Wiley.

Ballast, D. K. (1984) *The Architect's Handbook*, Englewood Cliffs, NJ, Prentice-Hall.

Ballast, D. K. (1986) *Practical Guide to Computer Applications for Architecture and Design*, Englewood Cliffs, NJ, Prentice-Hall.

Barrett, P. and Males, R. (eds) (1991) *Practice Management*, London, Spon.

Beaven, L. (1983) *Architect's Job Book (4th rev. edn.)*, London, RIBA.

Blau, J. R. (1984) *Architects and Firms: a Sociological Perspective on Architectural Practice*, Cambridge, MIT Press.

Brawne, M. (1983) *Arup Associates (1st edn.)*, London, Lund Humphries.

Burstein, D. (1982) *Project Management for the Design Professional*, New York, Whitney Library of Design, London, Architectural Press.

Case and Company (1968) *The Economics of Architectural Practice*, Washington, DC, American Institute of Architects.

Case and Company (1968) *Profit Planning in Architectural Practice*, Washington, DC, American Institute of Architects.

Clarke, D. (1994) *The Architecture of Alienation*, London, Transaction Publishers.

Coxe, W (1980) *Managing Architectural and Engineering Practice*, New York, Van Nostrand Reinhold.

Creswell, H. B. (1929) *The Honeywood File: an adventure in building*, London, Architectural Press.

Creswell, H. B. (1930) *The Honeywood Settlement: a continuation of The Honeywood File*, London, Architectural Press.

Crinson, M. and Lubbock, J. (1994) *Architecture, Art or Profession? Three hundred years of architectural education in Britain*, Manchester, Manchester University Press.

Cuff, D. (1991) *Architecture: the Story of Practice*, Cambridge, MA, MIT Press.

Cuff, D. (1991) *The Culture of Practice: Architecture in American Society*, Cambridge, MA, MIT Press.

de Wolfe, I. (1971) *Civilia*, London, Architectural Press.

Dibner, D. R. (1972) *Joint Ventures for Architects and Engineers*, New York, McGraw-Hill.

Dingwall, R. and Lewis, P. (eds) (1983) *The Sociology of the Professions*, Basingstoke, Macmillan.

Duffy, F., Cove, C. and Worthington, J. (1976) *Planning Office Space*, London, Architectural Press.

Durkheim, E. (G. Simpson trans.) (1993) *The Division of Labour in Society*, New York, Free Press.

Economic and Social Research Council (1994) *Research Specification for the ESRC Learning Society Knowledge and Skills for Employment Programme*. Swindon, ESRC.

Eley, J. and Symes, M. (1993) *Recent Changes in Architectural Practice*, occasional papers in architecture and urban design, No. 2, Manchester, University of Manchester School of Architecture.

Esher, L. (1981) *The Broken Wave*, London, Allen Lane.

Essex County Council (1973) *Design Guide for Residential Areas.*

Evans, B. H. (1969) *Architectural Programming: Emerging Techniques of Architectural Practice*, Washington, DC, AIA

Evans, N. (1981) *The Architect and The Computer*, London, RIBA.

Foote, F. (1978) *Running an Office for Fun and Profit: Business Techniques for Small Design Firms*, Shroudsberg, PA, Dowsen, Hutchinson and Ross.

Foxhall, W. B. (1974) *Techniques of Successful Practice for Architects and Engineers*, New York, McGraw-Hill.

Geertz, C. (1983) *Local Knowledge: Further Essays in Interpretive Anthropology*, New York, Basic Books.

Getz, L. (1984) *Financial Management for the Design Professional: a Handbook for Architects, Engineers and Interior Designers*, New York, Whitney Library of Design.

Golzen, G. (1984) *How Architects get Work*, London, Architectural Press.

Grant, D. P. (1983) The Small-Scale Master Builder, San Luis Obispo, CA, Small-Scale Master Builder.

Green, R. (1986) *The Architect's Guide to Running a Job (4th edn.)*, London, Architectural Press.

Greenstreet, B. (1984) *The Architect's Guide to Law and Practice*, New York, Van Nostrand Reinhold.

Gutman, R. (1988) *Architectural Practice: a Critical View*, Princeton, NJ, Princeton Architectural Press.

Guttridge, B. (1973) *Computers in Architectural Practice*, London, Cosby Lockwood Staples.

Harada, J. (1985) *The Lesson of Japanese Architecture*, New York, Dover.

Harvey, D. (1989) *The Condition of Post-Modernity*, Oxford, Blackwell.

Haviland, D. S. *Managing Architectural Projects: the Process*, Washington, DC, AIA.

Haviland, D. S. *Managing Architectural Projects: the Project Management Manual*, Washington, DC, AIA.

Holton, R. J. and Turner, B. S. (1993) *Max Weber on Economy and Society*, London, Routledge.

Jencks, C. (1977) *The Language of Post-Modern Architecture*, London, Academy.

Jencks, C. and Silver, N. (1972) *Adhocism: The Case for Improvisation*, London, Secker and Warburg.

Jenkins, F. I. (1961) *Architect and Patron: A survey of professional relations and practice in England from the sixteenth century to the present day*, London, Oxford University Press.

Kaderlan, N. S. (1991) *Designing Your Practice, A Principal's Guide to Creating and Managing a Design Practice*, New York, McGraw-Hill.

Karner, G. E. (1989) *Contracting Design Services*, Washington DC, American Society of Landscape Architects.

Kaye, B. (1960) *The Development of the Architectural Profession in Britain: A sociological study*, London, Allan and Unwin.

Larson, M. S. (1977) *The Rise of Professionalism: A sociological analysis*, Berkeley, University of California Press.

Leighton, N. L. (1984) *Computers in the Architectural Office*, New York, Van Nostrand Reinhold.

Lewis, J. R. (1978) *Architect's and Engineer's Office Practice Guide*, Englewood Cliffs, NJ, Prentice-Hall.

Little, S. (1988) *The Organisational Implications of Computer Technology for Professional Work*, Aldershot, Brookfield, VT, Avebury.

Lyall, S. (1980) *The State of British Architecture*, London, Architectural Press.

MacEwen, M. (1974) *Crisis in Architecture*, London, RIBA.

McCormick, B. J., Kitchin, P. D., Marshall, G.P., Sampson, A. A., Sedgwick, R. (1974) *Introducing Economics*, Harmondsworth, Penguin.

Mattox, R. F. *Financial Management for Architects*, Washington DC, AIA.

Middleton, M. (1969) *Group Practice in Design*, New York, G. Braziller.

Moulin, R., Dubost, F., Gras, A., Lautman, J., Martinon, J. P., Schapper, D. (1973) *Les Architectes: Metamorphose d'une profession liberale*, Paris, Callman-Levy.

Newman, O. (1973) *Defensible Space*, London, Architectural Press.

Oliver, P. *Architecture: an Invitation*, Oxford, Basil Blackwell.

Orr, F. (1982) *Professional Practice in Architecture*, New York, Van Nostrand Reinhold.

Perkin, H. (1989) *The Rise of Professional Society*, London, Routledge.

Piven, A. *Compensation Management: a Guideline for Small Firms*, Washington, DC, AIA.

Price, C., Newby, F. and Svan, R. H. (1971), *Air Structures*, London, HMSO.

Radford, A. (1987) *CAD Made Easy: a Comprehensive Guide for Architects and Designers*, New York, McGraw-Hill.

Rossman, W. E. (1972) *The Effective Architect*, Englewood Cliffs, NJ, Prentice-Hall.

Royal Institute of British Architects (1962) *The Architect and his Office*, London, RIBA.

Royal Institute of British Architects (1989) *RIBA Practices 89: A Directory of RIBA Practices*, London, RIBA.

Royal Institute of British Architects (1990, 1991, 1992) *Architects' Employment and Earnings*, London, RIBA.

Royal Institute of British Architects (1992) *Strategic Study of the Profession: Phase One: Strategic Overview*, London, RIBA.

Ryan, D. L. (1983) *Computer-Aided Architectural Graphics*, New York, Marcel Dekker.

Saint, A. (1983) *The Image of the Architect*, New Haven, Yale University Press.

Salaman, G. (1974) *Community and Occupation*, Cambridge, CUP.

Seidel, A. D. (1992) Breaking a myth of architecture education: effective management and effective design go hand-in-hand. In: Aristides, M. and Karaletsov, C. (eds) *Socio-Environmental Metamorphoses*, Thessaloniki, Aristotle University.

Seidel, A. D. (ed.) (1993) A knowledge base for the practice of architecture: results of a survey of practitioners in the United Kingdom. In: Spreckelmeyer, K. *et al. Knowledge-Based Architectural Education: Reconfiguring the Studio*, Washington, DC, Architecture Research Centers Consortium.

Seidel, A. D. (1993) A development profession searches for knowledge base: architects in the United Kingdom. In *People and Physical Environment Research*, Vol. 43, pp. 36–42.

Seidel, A. D. (1995) The knowledge needs architects request. In Seidel, A. D. (ed.) *Banking of Design?*, Oklahoma city, Environmental Design Research Association, pp. 18–24.

Senior, D. (1964) *Your Architect*, London, Hodder with RIBA.

Sharp, D. (1986) *The Business of Architectural Practice*, London, Collins.

Shoshkes, E (1989) *The Design Process*, New York, Whitney Library of Design.

Symes, M. (1990) The culture of British architects: 1968–1988. In: Pamir, H., Imamoglu and Teymur, N. (eds) *Culture, Space, History*, Ankara, Middle East Technical University, Vol. 5, pp. 72–84.

Symes, M. (1992) From responsibility to accountability: the British architectural profession and competition from others. In: Aristides, M. and Karaletsov, C. (eds) *Socio-Environmental Metamorphoses*, Thessaloniki, Aristotle University, pp. 177–79.

Taylor, N. (1971) *The Village in the City*, London, Temple Smith.

Thomsen, C. (1989) *Managing Brainpower: Organizing, Measuring Performance and Selling in Architecture, Engineering, and Construction Management Companies*, Washington, DC, AIA.

Turner, H. (1974) *Architectural Practice and Procedure (6th edn.)*, London, Batsford.

University of Arizona Press (1987) *Architectural Practice in Mexico City: a Manual for Journeymen Architects of the Eighteenth Century (English and Spanish)*, Tucson, UAP.

University of North Carolina Press (1990) *Architects and Builders in North Carolina*, Chapel Hill, NC, UNCP.

US Bureau of the Census (1982) *Census in Service Industries 1982*, Washington, DC, US Department of Commerce, Bureau of the Census.

Watkin, D. (1978) *Morality in Architecture*, Oxford, Clarendon.

Willis, A. J. (1981) *The Architect in Practice (6th edn. with minor rev.)*, London, New York, Granada.

Wilson, F. (1990) *Architecture: Fundamental Issues*, New York, Van Nostrand Reinhold.

Wolfe, T. (1981) *From Bauhaus to Our House*, New York, Farrar Straus Giroux.

Woodward, C. A. *Human Resources Management for Design Professionals*, Washington DC, AIA.

Yamey, B. S. (ed) (1973) *Economics of Industrial Structure*, Harmonds-worth, Penguin.

Yeomans, D. and Steel, M. (1994) *Professional Relationships and Technological Change in Great Britain*, report to Ministere du Logement, Plan Construction et Architecture, from the authors: University of Manchester School of Architecture.

Reference may also be made to apopropriate issues of:

> *Environment and Planning*
> *Journal of Architectural and Planning Research*
> *Journal of Architectural Education*
> *Journal of the Royal Institute of British Architects*
> *Progressive Architecture*
> *The Architects' Journal*